Military Transformation and the Defense Industry after Next

The Defense Industrial Implications of Network-Centric Warfare

Peter J. Dombrowski
Eugene Gholz
Andrew L. Ross

NAVAL WAR COLLEGE
686 Cushing Road
Newport, Rhode Island 02841-1207

Library of Congress Cataloging-in-Publication Data

Dombrowski, Peter J., 1963–
  Military transformation and the defense industry after next: the defense industrial implications of network-centric warfare / Peter J. Dombrowski, Eugene Gholz, Andrew L. Ross.
      p. cm.
  "Final Report, September 2002."
    ISBN 1-884733-24-7 (alk. Paper)
    1. Industrial mobilization—United States. 2. Defense industries—United States. 3. United States—Armed Forces—Reorganization. 4. United States. Navy—Procurement. I. Gholz, Eugene, 1971– II. Ross, Andrew L. III. Title.
    UA18.U5 D65 2002
    338.4'7355'00973—dc21
                                                        2002015464

Naval War College
Newport, Rhode Island
Center for Naval Warfare Studies
Newport Paper Number Eighteen
2003

President, Naval War College
Rear Admiral Ronald A. Route, U.S. Navy

Provost, Naval War College
Professor James F. Giblin, Jr.

Dean of Naval Warfare Studies
Professor Alberto R. Coll

Naval War College Press
Editor: Professor Catherine McArdle Kelleher
Managing Editor: Pelham G. Boyer
Associate Editor: Patricia A. Goodrich

Telephone: 401.841.2236
Fax: 401.841.3579
DSN exchange: 948
E-mail: press@nwc.navy.mil
Web: http://www.nwc.navy.mil/press

Printed in the United States of America

The Newport Papers are extended research projects that the Naval War College Press Editor, the Dean of Naval Warfare Studies, and the President of the Naval War College consider of particular interest to policy makers, scholars, and analysts.

The views expressed here are those of the authors and do not necessarily reflect those of the Naval War College, the Department of the Navy, or the Department of Defense.

Correspondence concerning The Newport Papers may be addressed to the Dean of Naval Warfare Studies. To request additional copies or subscription consideration, please direct inquiries to the President, Code 32A, Naval War College, 686 Cushing Road, Newport, RI 02841-1207.

Patricia A. Goodrich, Associate Editor, Naval War College Press, edits and prepares The Newport Papers.

ISSN 1544-6824

Contents

Figures and Tables	v
Foreword	vi
Executive Summary	vii
Acknowledgments	ix

INTRODUCTION: MILITARY TRANSFORMATION AND THE U.S. DEFENSE INDUSTRY ... 1

    The Defense Industrial Implications of Military Transformation ... 2

    Plan of Attack ... 3

THE NAVAL TRANSFORMATION CASE: NETWORK-CENTRIC WARFARE ... 5

    Enabling Elements ... 6

    NCW and Industry ... 10

TRANSFORMATION AND INNOVATION ... 13

    Sustaining and Disruptive Innovation ... 14

    The Customer Side of Innovation for Transformation ... 16

    Future Composition of the Defense Industrial Sector ... 19

THE EVOLVING DEFENSE INDUSTRIAL LANDSCAPE ... 21

    Consolidation ... 21

    Globalization ... 24

    Commercial-Military Integration ... 26

DEFENSE INDUSTRIAL SECTORS ... 29

    Shipbuilding ... 30

        The Shipbuilding Sector Today and Tomorrow ... 31

        NCW and Shipbuilding: New Performance Metrics? ... 34

        Customer-Supplier Relationships ... 42

        Sector Evaluation ... 44

Unmanned Vehicles	46
The UAV Sector Today and Tomorrow	47
NCW and UAVs	49
Emerging UAV Performance Metrics	49
Customer-Supplier Relationships	53
Sector Evaluation	55
Systems Integration	56
The System-of-Systems Integration Sector Today and Tomorrow	59
NCW and Systems Integration: Performance Metrics	64
Sector Evaluation	74

CONCLUSION   77

The Defense Industry and Military Transformation	77
Shipbuilding	79
UAVs	80
Systems Integration	81
Transforming the Navy	82

APPENDIX: GOVERNMENT AND NONGOVERNMENT INTERVIEWS   87

List of Abbreviations	91
Notes	95
Selected Bibliography	109
Selected Briefings	115
About the Authors	117
Titles in the Series	119

Figures and Tables

FIGURE 1	Project Focus I	5
FIGURE 2	Network-Centric Operations	7
FIGURE 3	The Information Grid—Detailed View	9
FIGURE 4	The EC5G: A Notional Depiction	10
FIGURE 5	Project Focus II	11
TABLE 1	U.S. Contractor Presence for Selected Military Platforms (1990–2000)	22
TABLE 2	U.S. Contractor Presence in Selected Military Product Areas (1990–2000)	22
TABLE 3	Major U.S. Private Shipbuilding Facilities—2001	33
TABLE 4	U.S. Private Sector UAV Manufacturers	48
TABLE 5	Examples of NCW-Related System-of-Systems Integration Organizations	59

Foreword

The first years of the new century see the American defense establishment in unabashed primacy but facing two serious challenges. The first is the transformation of the military itself and all of its attendant agencies no longer suited to both the possibilities and the constraints of an era without a peer competitor to set the march. The second is how, in a sea of new technologies, to craft a defense industrial base that both supports a transformed military and adapts to the dominant political and economic realities of the United States economy now embedded in a global trading system.

The three authors of this Newport Paper offer groundbreaking answers to the second challenge. Building on a series of unique interviews and refining their work through a chain of professional briefings, Peter Dombrowski, Eugene Goltz, and Andrew Ross argue persuasively that the new challenges are less new than predicted by transformation advocates, that there are perhaps more continuities and "old" interactions in the defense industrial base that produce efficiencies and effectiveness at levels that could not have been expected. Looking in detail at three sectors—shipbuilding, unmanned vehicles, and systems integration—they find that network-centric warfare requirements generate a range of defense industry implications but not the need for or the possibility of a complete overhaul.

Their analysis is impressive in its depth, but also in its reach. They integrate with new sophistication material drawn from practice and practitioners with cutting-edge business theory, especially Thomas Christensen's distinctions between disruptive and sustaining innovation and his emphasis on the variation in customer-supplier relationships. They are equally adept at the critical analysis of performance metrics, the industrial landscape of the present and future, and the process implications of proposed policy change.

The results of this research have been circulated among concerned decision makers over the last year. The Naval War College Press is pleased to bring this landmark work to the informed, attentive readership of the Newport Papers.

CATHERINE McARDLE KELLEHER
Editor, Naval War College Press

Executive Summary

Though still adjusting to the end of the Cold War, the defense industry is now confronted with the prospect of military transformation. Since the terrorist attacks on 11 September 2001, many firms have seen business improve in response to the subsequent large increase in the defense budget. But in the longer run, the defense sector's military customers intend to reinvent themselves for a future that may require the acquisition of unfamiliar weapons and support systems. Joint and service visions of the military after next raise serious questions that require the attention of the Defense Department's civilian and uniformed leadership and industry executives alike:

- What are the defense industrial implications of military transformation?
- Will military transformation lead to major changes in the composition of the defense industrial base?

This study employs *network-centric warfare,* a Navy transformation vision that is being adopted increasingly in the joint world as a vehicle for exploring the defense industrial implications of military transformation. We focus on three defense industrial sectors: shipbuilding, unmanned vehicles, and systems integration.

The transformation to NCW will require both sustaining and disruptive innovation—that is, innovation that improves performance measured by existing standards and innovation that defines new quality metrics for defense systems. The dominant type of innovation needed to support transformation varies across industrial sectors; some sectors face more sustaining than disruptive innovation, while some sectors will need more disruptive than sustaining innovation as they supply systems for the "Navy after Next."

Military transformation does not entail wholesale defense industrial transformation. In the systems integrations sector, much of the innovation required to effect network-centric warfare is likely to be sustaining rather than disruptive. In the parts of the defense industrial base that build platforms, on the other hand, the standards by which proposals are evaluated for the Navy after Next will be somewhat different than the standards used in the past. As a result, transformation could significantly change the industrial landscape of shipbuilding. The unmanned-vehicle sector falls somewhere in between; because unmanned vehicles have not been acquired in quantity in the past, their performance metrics are not well established. Existing suppliers of unmanned vehicles will have a role in the future industry, but some innovative concepts and technologies may come from nontraditional suppliers, such as start-up firms.

The U.S. Navy bears the responsibility of transforming itself. Internally, it must find ways to deconflict the needs of the current Navy and the "Next Navy" from the needs of the Navy after Next if industry is to support its long-term transformation requirements. External, pervasive organizational and political obstacles to transformation require that the Navy carefully manage its relationships with Congress and industry. Recognition that military transformation need not drive existing defense firms out of business will facilitate that task.

Acknowledgments

In late spring of 2000, Vice Admiral Arthur K. Cebrowski, U.S. Navy, then the president of the Naval War College and now director of the Defense Department's Office of Force Transformation, asked the College's Center for Naval Warfare Studies to undertake a defense industrial base study. Dr. Alberto Coll, Dean of Naval Warfare Studies, assigned the project to us in August 2000. Vice Admiral Cebrowski and Dean Coll enthusiastically embraced our proposal to explore the defense industrial implications of military transformation, particularly naval transformation, a subject near and dear to Vice Admiral Cebrowski. We thank both of them for their support, encouragement, and patience. We also thank Rear Admiral Rodney P. Rempt, U.S. Navy, who became president of the Naval War College in August 2001, for his support of our continuing work.

A large number of people contributed in a variety of ways to our work. Dr. Jonathan Pollack, the chairman of the Strategic Research Department, helped ensure that we had the resources necessary to complete this project. We would like to thank the more than 250 people in government, industry, and academia who took the time to meet with us during the course of our research for this project. Thomas P. M. Barnett, Michael Desch, Bradd Hayes, Hank Kamradt, Thomas Mahnken, Wayne Perras, and Harvey Sapolsky read all or parts of an earlier draft and provided valuable comments and suggestions. Lieutenant Commander William Murray, U.S. Navy, provided research and graphics support. Commander Marc Homan, U.S. Navy, also provided research support. Pauline Gagne and Cheryl Rielly provided critical administrative and graphics support.

This report was completed in September 2002. For further discussion of Navy transformation issues see Peter J. Dombrowski and Andrew L. Ross, "Transforming the Navy: Punching a Feather Bed?" *Naval War College Review* 56, no. 3 (Summer 2003), pp. 107–31.

Introduction
Military Transformation and the U.S. Defense Industry

The U.S. military is awash in visions of transformation. There is an array of joint and service visions of what has become known as the "military after next."[1] The rhetoric, if not yet the reality, of "revolution"—i.e., the revolution in military affairs (RMA) and, somewhat less radically, of "transformation"—is ubiquitous. While itself still adjusting to the end of the Cold War, the defense industry now confronts a customer that intends to reinvent itself for the future. On the heels of the struggle to consolidate recently merged assets, trim high debt-equity ratios left over from the 1990s wave of mergers, and respond to profit pressures from the post–Cold War decline in the defense budget, executives must deal with the new specter of military transformation.[2] While industry executives focus on the implications of transformation for the future of their firms, defense planners must ask whether the existing defense industrial sector is adequately prepared to support their visions of the military after next.

Joint Vision 2020, like *Joint Vision 2010* preceding it, foresees a military that dominates the full spectrum of military operations, from low-intensity conflicts to major theater wars, in new ways. Information superiority is to be the source of "dominant maneuver," "precision engagement," "focused logistics," and "full dimensional protection."[3] The Army's transformation project, complete with "Vision," "Force XXI," and "Army after Next," is billed as the most significant change for the service since World War I. The Army promises to deliver an "Objective Force" that will be responsive, deployable, agile, versatile, lethal, survivable, and sustainable.[4] The Air Force, which, like the Army, has belatedly discovered that it must be an expeditionary force,[5] promises in its own *Vision 2020* to deliver "Global Vigilance, Reach and Power" by fielding a "full-spectrum" aerospace force—that is, one designed to control and exploit not only the air but space as well.[6] The proposed force "encompasses aerospace capabilities to find, fix, assess, track, target, and engage any object of military significance on or above the surface of the Earth in near real time."[7] As for the Navy, network-centric warfare (NCW), advertised as a vision of warfare for the information age, is to guide, along with SEA POWER 21, the transformation of today's Navy into the Navy after Next. Resting upon the "supporting concepts" of information and knowledge advantage, "assured battlespace access," "effects-based operations," and forward sea-based forces, the Navy's exploitation of information technologies is to result in a "shift from platform-centric operations to Network-Centric Operations."[8]

Some have presumed, as it is tempting to do, that the new information-centric forces and doctrines will lead to a shift in military buying patterns so fundamental as to reorder the defense industrial landscape, with information technology firms assuming a heretofore unknown prominence. These analysts note that military leaders, in their revolutionary visions, are looking for ways to apply the tremendous advances in commercial information technology, highly visible in the "New Economy" of the 1990s, to military missions. In this view, because the defense sector's product cycle cannot keep pace with commercial information technologies, military transformation is likely to require defense industrial transformation.

This report, which analyzes the defense industrial implications of military transformation, takes a somewhat different tack. Drawing on well-known models of innovation, we develop a new framework that clearly specifies, arguing from core principles, what types of firms—established defense suppliers, established commercially oriented firms, or start-ups—are most capable of supporting transformation. Surprisingly, in view of the obvious technical capabilities of commercial information-technology firms, we find that current defense-oriented suppliers as a whole are likely to dominate the IT segment of the future defense market; it is the current defense-focused suppliers of large platforms (e.g., shipyards) that may be most vulnerable. Such firms are more likely than IT-oriented systems integrators to face competition from companies that now sell mostly to commercial customers or foreign navies. We conclude with policy and organizational recommendations for the military services and acquisition community that will help smooth the transformation process in the face of political opposition, budgetary constraints, and pressures for technological overreach.

The Defense Industrial Implications of Military Transformation

The military's declared intent to remake itself, and the Bush administration's oft-stated commitment to military transformation, pose the prospect of continued post–Cold War defense industrial disruption.[9] While analysts have begun to address the technological implications of transformation,[10] its defense industrial implications have not yet been systematically examined. Joint and service visions of the military after next raise serious questions that require the attention of the Defense Department's civilian and uniformed leadership and industry executives alike:

- What are the defense industrial implications of military transformation?
- Does military transformation require defense industrial transformation?
- Are traditional defense suppliers more likely to support a revolutionary or evolutionary approach to transformation, or will they resist all forms of transformation?
- Will traditional or nontraditional suppliers prove to be the richest sources of innovation?
- What kind of relationship between public-sector customers and private-sector suppliers might best facilitate transformation?
- How can the civilian and military leadership of the Department of Defense ensure that industry can, and will, support transformation?

If defense planners are serious about effecting military transformation, it is imperative that these questions receive attention at the outset.[11] Transformation is a process. Industry has a critical role to play in that process. Effective implementation of joint and service transformation visions requires that planners and programmers devote the necessary attention and resources to the technological and industrial dimensions of implementation. False starts down transformation paths that turn out to be technically or industrially impractical will prove costly. Time, money, and political capital are scarce resources. Given the needs for spending on current operations and near-term modernization, front-end transformation requirements and programs need to be carefully thought out. The sooner that defense planners come to grips with industry's role in military transformation, the better.

Plan of Attack

The questions and issues identified here are addressed in the five sections of this report. First, we identify the key technological trajectories along which the defense industrial base will have to develop and produce equipment for the Navy after Next by describing network-centric warfare. Second, we draw upon the literature on innovation to distinguish between sustaining and disruptive innovation and to develop a framework for identifying the types of firms capable of providing transformational goods and services to the military. Third, we lay the foundation for our analysis of specific defense industrial sectors by examining the major trends in the contemporary defense industrial landscape. Fourth, we present case studies of three defense industrial sectors whose products span the range of transformation requirements: shipbuilding, unmanned vehicles, and systems integration. We conclude with policy recommendations for ensuring that the relationship between the Navy (and by extension the military as a whole) and the U.S. defense industry is fruitful during the upcoming period of transformation.

The Naval Transformation Case
Network-Centric Warfare

An across-the-board examination of the defense industrial implications of *Joint Vision 2020* and Air Force, Army, and Navy visions of transformation would have been unmanageable. Feasibility and practicality dictated that we narrow the focus of this project. That was done in two ways. First, as illustrated in figure 1, we narrowed our focus from joint and service visions of military transformation generally to naval transformation specifically. Second, even while exploring the "big picture" defense industrial implications of military transformation, we focused on three industrial sectors—shipbuilding, unmanned vehicles, and systems integration—which, as explained more fully below, will have a critical role to play in naval transformation.

FIGURE 1
Project Focus I

MILITARY TRANSFORMATION
↓
JOINT AND SERVICE VISIONS
↓
THE NAVAL VISION:
NETWORK-CENTRIC WARFARE
↓
DEFENSE INDUSTRY AFTER NEXT

We use the case of naval transformation, as envisioned in the concept of NCW, as a means for exploring the defense industrial implications of military transformation. Network-centric warfare provides an ideal vehicle for our study. NCW is inherently joint; the Navy cannot implement it in isolation from the other services. At least as much as other service visions, NCW is broadly representative of military transformation, a naval manifestation of a more general phenomenon. Along with *Joint Vision 2020* and the visions of the other services, NCW emphasizes the need to bring the U.S. military into the information age. NCW envisions that new commercial technologies are to be applied to military tasks. Information technology is central to the entire transformation enterprise; frequently characterized as an "IT-RMA," it enables the realization of such prized capabilities as precision strike and a "common operational picture." Using NCW as the visionary touchstone and point of departure, therefore, will help us understand the defense industrial implications of not only naval transformation but also military transformation generally. By providing direct links to other service transformation visions, the three industrial sectors on which we focus facilitate a more universally applicable exploration.

The proponents of network-centric warfare portray it as an emerging vision of the future of war. That vision is driven by a particular understanding of the transformation of modern society from the industrial age to a postindustrial, or information, age at the beginning of the twenty-first century.[12] Advances in information technologies that have resulted in widespread socioeconomic changes will also revolutionize the conduct, if not the nature, of war.[13] In particular, the increasing use of networks for organizing human activities is touted as a means of reshaping the way American forces train, organize, arm, and fight.[14]

In brief, networks harness the power of geographically dispersed nodes (whether personal computers, delivery trucks, or warships) by linking them together into networks (such as the World Wide Web) that allow for the extremely rapid, high-volume transmission of digitized data (multimedia). Networking has the potential to increase exponentially the capabilities of individual nodes or groups of nodes and to facilitate the efficient use of resources. When networked, individual nodes gain access not only to their own resident capabilities but also, more importantly, to capabilities distributed across the network. The loss of a networked node need not be crippling; in a robust network its functions can and will be assumed by other nodes. Since networked nodes can share information efficiently, they can be designed as simple, low-cost adjuncts to the network itself.[15]

Enabling Elements

The U.S. armed forces are developing, initially by serendipity but increasingly by design, the capabilities necessary for network-centric operations (NCO).[16] In a draft of a capstone concept paper, the Navy Warfare Development Center (NWDC) identified four NCO "pillars," or supporting concepts: information and knowledge advantage, effects-based operations, assured access, and forward sea-based forces (see figure 2).[17]

FIGURE 2
Source: Navy Warfare Development Command, Network-Centric Operations: A Capstone Concept for Naval Operations in the Information Age, *"Executive Summary" (Newport, RI: Navy Warfare Development Command, draft dated 6/19/01), p. ii.*

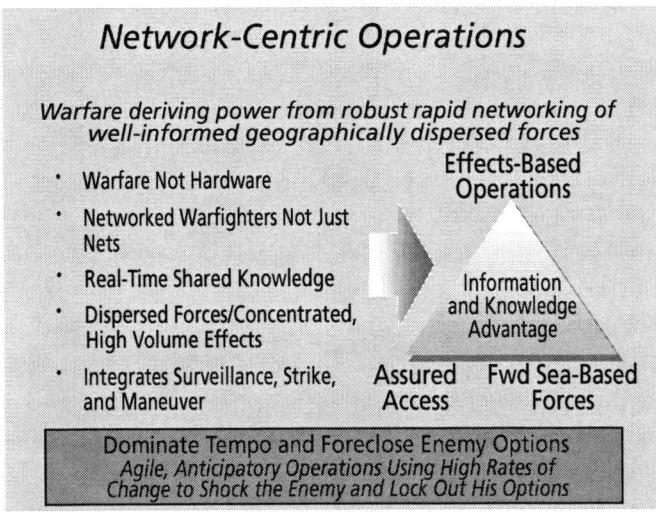

The postulated benefits of NCO provided by the pillars of information and knowledge advantage[18] and by effects-based operations[19] (EBO) include speed of command, self-synchronization, advanced targeting, and greater tactical stability. Netted sensors are to provide shooters and commanders with "unmatched awareness of the battlespace."[20] Within the battlespace, warfighters are to be able to "self-synchronize" their activities to accomplish a commander's intent by drawing upon a shared "rule set—or doctrine"[21]—and a "common operational picture" (COP). In essence, self-synchronization is accomplished by devolving decision making downward to the lowest appropriate level, thus allowing warfighters to respond directly and quickly to tactical, operational, and even strategic challenges. Fires are to be employed for effects-based operations rather than attrition-based warfare. Precision-guided munitions, in conjunction with advanced intelligence, surveillance, and reconnaissance (ISR) capabilities will allow targets to be hit with greater economy—simultaneously, rather than sequentially—greatly increasing the possibility of inflicting disproportionate effects, particularly psychological, on the adversary. Tactical operations may thus achieve strategic objectives.

Finally, by geographically dispersing sensors, shooters, and their supporting infrastructure within an overarching network, U.S. forces will be able to achieve greater tactical stability—a favorable balance between survivability and combat power.[22] Fires, rather than forces, will be massed, and they will be delivered from beyond visual range. Ideally, EBO, fueled by information and knowledge superiority, will enable U.S. forces to "lock in success and lock out enemy solutions" and options.[23] Smaller, lighter, faster, less complex, and less expensive nodes (i.e., platforms) linked by interoperable, highly redundant, self-healing networks will present adversaries with fewer high-value targets and improve the robustness of operations against a determined foe.

Implicitly at least, NCO is a joint vision that harnesses capabilities from all services; it is applicable to warfare on land, air, and sea.[24] That NCO is a Navy concept with naval origins, however, is evident in the two pillars that are distinctly naval: assured access and forward-deployed sea forces. "Assured access"[25] refers to the ability of the U.S. armed forces to gain entry to and use both overseas infrastructure, such as ports and airfields, and the battlespace itself, even when confronted with a capable and hostile adversary.[26] No sanctuary is to be ceded to the adversary. It is the job of the Navy and the Marine Corps to enable and ensure access for follow-on forces from the Air Force and the Army—the heavier forces necessary to fight and win major regional contingencies. The Navy accomplishes this through the combat capabilities inherent in its forward-deployed presence assets (i.e., the ability to operate in the littoral).[27] Since sea-based forces "do not rely on permissive access to foreign shore installations that may be withdrawn or curtailed," they "furnish an assured infrastructure for additional joint forces."[28]

In its Capabilities of the Navy after Next (CNAN) project, NWDC has sought to determine what technologies, weapons, platforms, and systems the fleet requires to enable it to conduct NCO. The principal "enabling element" of NCO is a set of information, sensor, and engagement grids capable of linking all nodes of the network with each other and with the wider "information backplane"—the World Wide Web and DoD-specific networks. NCO relies greatly on the development and deployment of large numbers of capable sensors to populate the sensor grid and provide a COP; sensors are to be ubiquitous. This is not a network but a network of networks, "a global grid of multiple, interoperable, overlapping sensor, engagement, and command nets."[29]

Among existing programs, as illustrated in figure 3, the Cooperative Engagement Capability (CEC), IT-21, the Radar Modernization Program (RMP), the Web Centric Anti-Submarine Warfare Net (WeCAN), and the Navy–Marine Corps Intranet (NMCI) will help the Navy evolve further toward the ability to conduct NCO.[30] According to the NWDC, a critical future step is the deployment of the multitiered—space, air, surface/ground and undersea—Expeditionary Sensor Grid (ESG), combining, among other things, invasive sensing systems, unmanned platforms, massively distributed information systems, and computer-network attack and defense capabilities.[31] At its simplest, the ESG is a "toolbox of sensors and networks necessary to build . . . real-time battle-space awareness."[32]

FIGURE 3
The Information Grid—Detailed View
Source: http://spica.or.nps.navy.mil/netusw/CebrowskiNetWar/sld005.htm.

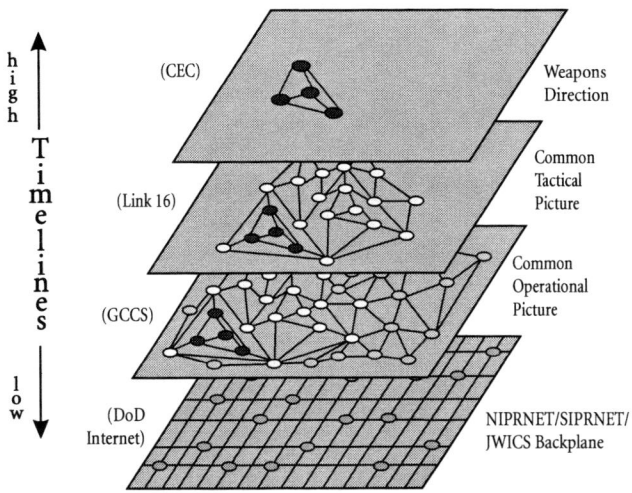

The most robust form of NCW also features smaller, lighter, faster, less complex, and less expensive platforms (nodes) that will facilitate self-synchronization, swarming tactics, and greater tactical survivability. Prominently featured in this array of innovative nodes are unmanned vehicles that will deploy sensors throughout the future battlespace or serve as sensors, communications relays, or weapons platforms. Perhaps the most significant platform-related question from a naval standpoint is whether NCW requires such innovative design concepts as small littoral combatants (formerly known as STREETFIGHTER), fast lift, and small-deck aircraft carriers. Complexity is to reside in the web rather than in the node; the complex, expensive platform nodes that populate the traditional, or "legacy" force will be displaced by simpler, less expensive ones. In today's Navy, existing platforms are being networked via, for instance, CEC and IT-21. In the future's network-centric Navy, nodes will be tailored to network requirements from their earliest conception.

In the spring of 2002, "FORCEnet," as portrayed in the "Naval Transformation Roadmap," emerged as the Navy's framework for implementing NCW.[33] Originally developed by the CNO's Strategic Studies Group, FORCEnet is billed variously as putting the "warfare" in network-centric warfare and as "the next generation of NCW." It is intended to provide the architecture for integrating NCW components: network systems, sensors, decision aids, weapons, platforms, people, and infrastructure. FORCEnet is to conceptually and physically network SEA POWER 21's capabilities—offensive (Sea Strike), defensive (Sea Shield), and "persistent presence" (Sea Basing).[34] It serves as an umbrella for existing programs such as NMCI, IT-21, CEC, and NFN (Naval Fires

Network) as well as for such major future programs as the Expeditionary Command and Control, Communications, Computers, and Combat Systems Grid (EC5G) and the ESG (see figure 4).[35] With the promulgation of SEA POWER 21, FORCEnet, and the Naval Transformation Roadmap, network-centric concepts have been firmly embedded in official statements on naval transformation.

FIGURE 4
The EC5G: A Notional Depiction
Source: https://ucso2.hq.navy.mil/n7/webbas01.nsf/(vwwebpage)/webbase.htm?OpenDocument

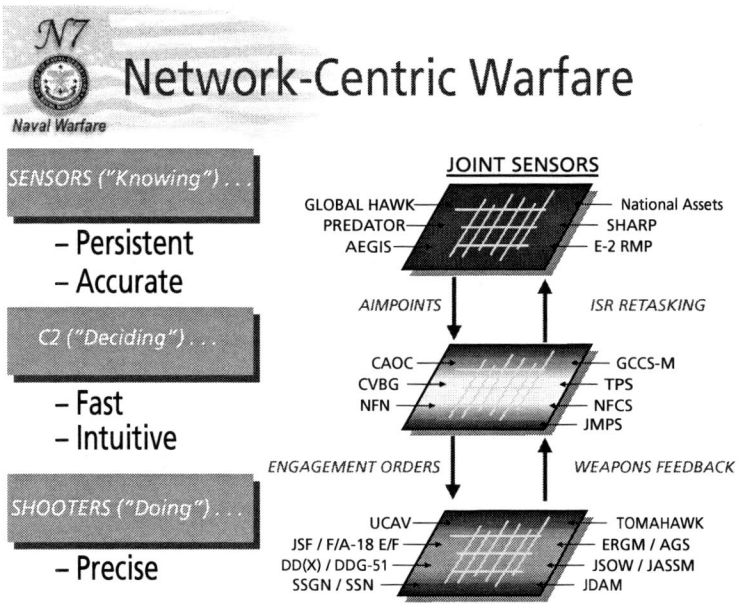

NCW and Industry

Our exploration of the defense industrial implications of network-centric warfare focuses, as illustrated in figure 5, on three defense industrial sectors: shipbuilding, unmanned vehicles, and systems integration. These sectors were selected for three reasons. First, they span the network and node components of NCW. Second, the shipbuilding sector is unique to the naval case,[36] whereas unmanned vehicles and systems integration are common to joint and service visions. Third, the potential role of commercial information technologies is prominent in all three sectors.

The shipbuilding sector, in its present form or an altered one, will bear the burden of designing, building, and supporting the transformational naval platforms envisioned by NCW architects. Unmanned vehicles, unlike naval platforms, are a shared feature of transformation visions. Aerial, ground/surface, and undersea unmanned vehicles are

FIGURE 5
Project Focus II

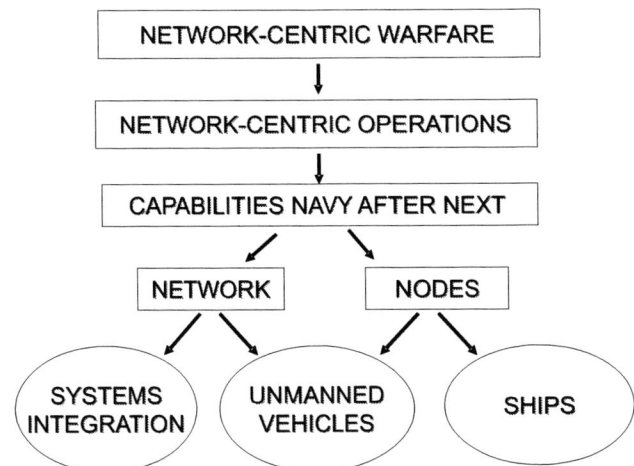

envisioned as network nodes that can be employed as sensor and weapons platforms, sensor distributors, and as communications relays that connect new and legacy manned platforms to the network.

Integrating NCW's nodes or platforms—manned and unmanned, legacy and next generation—and its envisioned information, sensor, and engagement grids to create a network of networks, or system of systems,[37] presents formidable challenges. The most significant benefits of the network will not be realized unless its myriad components are designed to optimize their relationship to the network—exporting some functions to other parts of the network while maintaining the internal capabilities necessary to self-synchronize and operate in a coordinated, decentralized fashion. Furthermore, the integration of a system of systems is not only critical for naval transformation but is a shared transformation requirement. Neither NCW nor *Joint Vision 2020* or the other service visions can be realized without overcoming significant systems-integration challenges. Transformation's demanding systems-integration requirements have been recognized by Kent Kresa, the former chairman of the Northrop Grumman Corporation: "Before us is a future requiring advanced computer processing power, global networks, a wide range of integrated satellite surveillance sensors and a growing inventory of effective and relatively inexpensive precision munitions powered by precise information. . . . But while we know these things will be needed, often we do not know how to integrate them into a cohesive military force."[38] Similarly, within the Navy, Rear Admiral Robert G. Sprigg, former Commander, Navy Warfare Development Command, has emphasized the challenge of developing an integrated architecture for NCO's expeditionary sensor grid, with its array of space, air, surface/ground, undersea, and cyberspace sensors, "that can handle this merge of thousands of inputs."[39] As our examination of the systems integration sector will demonstrate, however, there is reason to believe that established systems integrators possess the capabilities needed for system of systems integration.

Harnessing information technologies is at the heart of the emerging revolution in military affairs and the transformation process that will implement the new visions of warfare. Indeed, the emerging revolution in military affairs has been characterized as an "IT-RMA." The role of commercial IT in the evolving transformation parallels that of commercial technologies in the nineteenth century's industrialization of warfare. The sources from which the military will draw its revolutionary information technologies constitute an important aspect of this study's examination of the shipbuilding, unmanned vehicle, and systems-integrations sectors.

Transformation and Innovation

The military's transformation proposals envision substantial force structure, doctrinal, and organizational innovations within the services, and technological innovations in the goods and services that the military purchases from the defense industry. In particular, the Navy's preparations to implement network-centric warfare as it constructs the Navy after Next raise several key questions regarding the relationship between a public-sector customer apparently intent on transformation and its private-sector suppliers of goods and services. Is the existing defense industrial base the best source of supply for the necessary equipment, or will the defense acquisition community need to reach out to new suppliers? Will existing suppliers have to transform themselves in response to the requirements for naval transformation? Does transformation require a new relationship between service customers and industry suppliers?

Former Under Secretary of Defense Pete Aldridge, Jr., has warned industry leaders, "You all have your work cut out for you."[40] But how disruptive will that work be for industry? During our meetings with them over the course of this project, private and public sector defense executives expressed varying views about the defense industrial implications of military transformation. Although their use of "disruptive" is broader and somewhat more intuitive than the more narrowly technical sense in which we employ the term in our discussion of innovation, it is clear that defense industry executives are attempting to work through the implications of military transformation for their businesses—and that they do not yet know precisely what those implications are. When asked whether military transformation requires defense industrial transformation and whether innovation that is disruptive for the military would be disruptive for industry as well, their responses spanned the spectrum.

A Raytheon executive responded with an unequivocal "Yes" when asked if military transformation required industrial transformation.[41] One former DoD executive with responsibility for defense industrial policies stated that industrial transformation and disruption were inevitable.[42] Another former high-level DoD executive, whose portfolio included defense industrial issues, argued that transformation will result in defense industrial restructuring and that industry would resist that restructuring.[43] A Northrop Grumman executive replied that he could tell the story either way; the story he in fact told was about military transformation absent defense industrial transformation and about military disruption absent industrial disruption.[44] A Boeing analyst argued that military transformation would be less disruptive for industry than for the military.[45] Electric Boat executives boasted that while the military, and particularly the Navy, have talked about transformation, Electric Boat actually transformed itself during the 1990s.[46]

Our attempt to address the critical questions with which industry executives and defense planners alike are grappling begins with a discussion of key concepts from the literature on innovation. Those concepts help explain the linkages between customer-supplier relationships and innovation. Essentially, different kinds of innovations tend to be developed by firms in different relationships with their customers. Military transformation calls for particular types of innovations in each defense industrial sector. We use innovation theory to explain what kinds of firms within each sector can be expected to supply the products required to implement transformation. In the subsequent sections on the shipbuilding, unmanned vehicles, and systems integration sectors, we describe the types of innovation that transformation demands of suppliers, and we consider the likelihood that transformation will require changes in the composition of the sector.

Sustaining and Disruptive Innovation

In using the literature on innovation, we address a specific, unusual question: can an established customer-supplier relationship (such as that which exists between the Navy and the defense industry) generate innovative products? This question is not addressed by most theories of innovation, military or commercial. Most research emphasizes the challenges of creating new technological concepts (who thinks of innovations?) and of adapting organizations to capitalize on new technologies (how do inventions become usable products?).[47] Recent work by Clayton Christensen, however, offers a way to explore the potential rise and fall of leading firms in the existing supplier base, using the distinction between sustaining and disruptive innovations.[48] The key insight is that firms with established customer relationships are very good at producing sustaining innovations but that those same firms will not be (or are not inclined to be) interested in disruptive innovations. According to Christensen, disruptive innovations generally require new suppliers, dealing with new customers.

Sustaining innovations build on familiar product-quality metrics and customer-supplier relationships—"What all sustaining technologies have in common is that they improve the performance of established products along the dimensions of performance that mainstream customers in major markets have historically valued."[49] Sustaining innovations, no matter how complex, technically radical, or resource intensive, almost never drive established firms out of business; instead, they tend to reinforce the success of current suppliers. Expert technical and financial advisors to both suppliers and customers predict that sustaining innovations will prove feasible, and they understand how to update strategic plans to capitalize on innovation. Customers and suppliers can then cooperate on defining the technical and market requirements to develop the new product.[50] Military transformation is likely to reinforce the role of established members of the defense industrial base in those sectors in which it demands sustaining innovations.

Disruptive innovations, on the other hand, often perform less well at first, measured by the traditional standards, but they introduce new metrics that appeal to a different customer base. Of course, not all new technologies that perform poorly qualify as disruptive innovations; they must establish a trajectory of rapid performance improvement that, building on experience gained in fringe or niche markets, overtakes the performance of the old market-leading product on traditional measures of performance.[51] Unfortunately, it is especially difficult to predict improvements upon previously unrecognized product attributes. The standard operating procedures of established firms' strategic planning departments, based on sound models developed by technical and financial experts, will tend to weed out highly uncertain investments that hold the potential to yield disruptive innovations. Business strategists fear that new technologies will develop into "bad performers" in the long run rather than revolutionary products that fundamentally change the market.[52] Existing firms' biggest customers, with whom they naturally maintain close relationships, also shun the risk of inferior performance.[53] Consequently, it is new firms—lacking standard operating procedures or well-developed customer relationships—that are most likely to invest in disruptive innovations. When a start-up firm's investment succeeds, the industrial landscape is transformed, as the start-up replaces the pre-innovation market leaders. Military transformation is more likely to result in new suppliers making an appearance in the defense marketplace when it requires disruptive innovation.

Christensen's analysis of the dynamics of customer-supplier relationships and innovation must be applied with care to the defense sector. Transformation differs from the usual case in which a customer decides whether to accept or reject an innovation offered by a nontraditional, upstart supplier. After all, in the defense sector, the demand for transformation and innovation, whether sustaining or disruptive, originates with the customer. That "demand pull" weakens the usual dynamic in which established firms decline to participate in disruptive innovations.

Moreover, the unique characteristics of the defense industry may alter the traditional entrepreneurial route by which disruptive innovations usually drive established suppliers out of business, because the factors that drive military acquisition decisions are unfamiliar to most business executives. In the commercial world, entrepreneurs and venture capitalists generally understand the manufacturing, marketing, and ultimately profit-making rationales of their customers in adopting niche markets early. For the defense industry, however, customers' operational concerns depend on military concepts that are alien to most technological entrepreneurs. Defense acquisition projects require entrepreneurs to understand and exploit unfamiliar, noneconomic strategies in their business plans. As a result, even in sectors of the defense industry where transformation introduces new performance metrics—sectors in which the mechanistic application of Christensen's theory would suggest that the established firms are vulnerable to new entrants—established defense firms may play a crucial role as brokers between entrepreneurs and military customers. Joint ventures with, or acquisitions of, start-ups by traditional defense contractors will enable newcomers to translate more readily the language of military operations in which military doctrine developers express their professional expertise into technological and industrial requirements.[54] Solid relationships between established defense firms and protransformation customers will facilitate communication that speeds investment on the new technological trajectory.

The extent to which the existing defense industrial base is positioned to support military transformation will vary across sectors of the business. That variation will depend on the extent to which the innovations required from that sector are sustaining or disruptive and on the extent to which the existing, trusted relationship between defense firms and their military customers is necessary to broker the requirements definition and project management processes.

The Customer Side of Innovation for Transformation

Even though customer resistance to disruptive innovation is reduced in the case of military transformation by the customer's commitment to a vision of future, information-intensive warfare, some customer resistance remains. The customer-supplier dynamic here cannot be reduced to either customer comfort with sustaining innovation provided by established firms or to customer resistance to disruptive innovation offered by upstart suppliers. Customer resistance in this case is a response not to innovation originating in industry but to conflict within the military itself about the future of warfare. The military is not a single, unified customer; each service promotes a different vision.[55] In fact, various communities within individual services (for instance, the Navy's three "baronies"—surface, subsurface, and aviation) compete for roles, missions, and resources. Some services and communities are more committed to transformation than others; and the most committed have emerged as, in effect, the functional equivalent of Christensen's niche customers, early adopters of potentially disruptive innovations. These transformation advocates aspire to become *the* customers for the defense industry.

The military services develop new doctrines and capabilities in reaction to the changing strategic environment and to lessons learned from military operations and wars. Since the early 1990s, civilian and military defense planners have argued that the Cold War's end requires a capabilities-based rather than a threat-based approach.[56] Military analysts continue to debate the mechanisms by which the services develop innovative doctrines and capabilities. Three prominent theories, developed by Barry Posen, Steven Rosen, and Owen Cote, respectively, suggest that the current uncertainty about the future of warfare is a normal stage in the process of military innovation.[57]

Barry Posen argues that most innovation in military doctrine stems from the actions of civilian politicians. In response to the changing goals of the state or to changes in the international political-military environment, civilian leaders revisit the country's grand strategy. In time of high external threat, civilian leaders can intervene to disrupt the standard operating procedures by which the military services would otherwise continue to invest in training and equipment to serve the previous grand-strategic goals.[58] In the context of the contemporary transformation debate, Posen's theory is exemplified by the Bush administration's commitment to changing the face of the American military and by Secretary of Defense Donald Rumsfeld's very public efforts to reassert the primacy of the civilian Office of the Secretary of Defense over the uniformed military.[59] On the other hand, the absence of a traditional security threat to the United States from a "peer competitor" may allow the civilian national security agenda to be dominated by pork-barrel concerns that will not drive the services toward long-term doctrinal innovation. The short-term pressures of the war on terrorism may consume the civilian (and military) leadership, and prospects for military innovation may fade. Without doctrinal innovation, pressures for technological innovation in the defense industry will recede as well.[60]

In contrast to Posen, with his emphasis on external pressures, Stephen Rosen explains military innovation by focusing on the internal dynamics of military organizations. Rosen argues that peacetime military innovations depend on visionary officers who work steadily to solve problems with existing strategic and operational concepts that they identify through their expertise and operational experience. Military innovation succeeds when high-ranking visionaries protect creative junior officers from political threats and when those junior officers can gain promotion on the basis of their innovative ideas.[61] Vice Admiral Arthur Cebrowski's vital role in developing the core concepts of network-centric warfare looks like an example of Rosen's innovation mechanism at work. Furthermore, the establishment of the Navy Warfare Development Command, now the "organizational home" for thinking about network-centric warfare, may provide the key institutional support for naval transformation. On the other hand, Cebrowski's retirement and appointment as the first director of the Pentagon's civilian-led Office of Force Transformation might undercut his ability to protect a protransformation promotion path within the Navy. Moreover, the increasing involvement of commanders of the unified regional and functional combatant commands in preparing service acquisition plans and budget proposals is introducing an institutional bias toward current operational concerns rather than toward long-term doctrine development and future modernization. According to Rosen's logic, this trend presents a threat to military transformation.

Owen Cote traces military innovation to inter- and intraservice rivalry for roles and missions adjudicated by the civilian leadership; his theory envisions a prominent role for both civilian and military leaders.[62] Leaders of warfighting communities—such as, in the Navy, surface warfare officers, submariners, and aviators, and in the wider interservice context, Army and Marine infantry officers—gain professional status when they can offer the National Command Authority the best solutions to particular strategic or operational problems.[63] Each community can also offer its military judgment to discredit competing proposals, and each may draw technical advisors into the process to support its own proposals or to undercut alternatives. According to Cote's theory, the best innovative doctrines are adopted through the traditional American process of pluralism and open debate.[64] In this view, Admiral Cebrowski's move to the Office of Force Transformation can be seen as an endorsement of the evolving network-centric warfare vision by the civilian leadership. Full implementation of the vision, however, will still require a sustained commitment of resources from political leaders and a willingness to choose among competing military transformation options. And the Office of Force Transformation would need to be appropriately and fully staffed and given a role in the development of the Future Years Defense Plan and the Defense Planning Guidance (and the new Transformation Planning Guidance).

If the United States decides to commit itself to military transformation, through whatever process of doctrinal innovation, the resulting vision of future warfare will produce a new set of equipment requirements. Ultimately, most technological innovation in the defense industry comes from firms responding to new requirements derived from doctrine, although other influences, including the political economy and the political geography of weapons production, may distort the outcome.[65] The requirements for new defense systems will presume the success of certain technological innovations—some of which will be sustaining and some of which will be disruptive—that will in turn shape the likely future defense industrial landscape. Yet because new military doctrines depend on certain technological innovations, feedback from expert technical advisors—through contact with civilians, military doctrine development commands, and military acquisition organizations—should contribute to debates about the future of warfare.

Future Composition of the Defense Industrial Sector

Combining lessons from the business and military analysis literatures on innovation gives us a framework for determining which types of firms—established defense contractors, leading commercial information technology firms, or small start-up ventures—will populate each sector of the future defense industry.[66] The distinction between sustaining and disruptive innovation has significant implications for military transformation generally and for the transition to network-centric warfare specifically. First, NCW's requirements for sustaining and disruptive innovation will determine whether established, traditional defense suppliers or nontraditional suppliers, particularly commercial IT firms and start-ups, are best positioned to support naval transformation. Since the requirement for sustaining and disruptive innovation appears to vary across defense industrial sectors, the opportunities for nontraditional suppliers will vary across sectors as well. Second, because the services' technical and acquisition organizations—the defense industry's key customers—will exert a tremendous influence on the trajectory of technological change in the defense sector, the management of the customer-supplier relationship throughout the systems development process will be central to efforts to prepare the defense industry to implement transformation.

The Evolving Defense Industrial Landscape

Analyzing the defense industry of today in order to understand the defense industry of the future is inherently risky. After all, if analysts had sought to discern the contours of the industrial landscape of the year 2000 using data from 1990, they would have been wide of the mark. With the Soviet Union yet to collapse and a great deal of uncertainty remaining over the fate of the transitional countries in Central Europe, few would have predicted the large decreases in the U.S. defense budget or industry's struggle to remain viable in the face of declining markets. That said, doing the opposite—attempting to peer into the future without assessing the current environment—would be equally foolhardy. As political economists argue, the future is path dependent; where you are going depends on where you are and the choices you made in the past.

In this section we lay a foundation for discussing the future of the defense industry, first discussing how the defense industrial landscape arrived at its current state. We examine several of the prominent themes present in most studies of the defense industry—consolidation, globalization, and commercial-military integration.

Consolidation

Defense industrial consolidation refers to an ongoing process of mergers and acquisitions that have transformed the defense industrial landscape.[67] Consolidation has dramatically altered the defense industrial landscape. As Jeffrey Bialos, a former Deputy Under Secretary of Defense for Industrial Affairs, has pointed out, "What were 33 separate businesses in 1990 are 5 large defense firms today [2000]."[68] The number of separate businesses plunged in many sectors of the defense industry during the 1990s (see tables 1 and 2). Many of the most famous names in American industry, from General Motors and Ford to Hughes Aircraft and McDonnell Douglas, have either left the defense business or exist today only as divisions of larger enterprises. The few remaining big defense firms generally comprise several formerly independent companies or defenseoriented divisions sold by other companies that have themselves left the defense business.

TABLE 1
U.S. Contractor Presence for Selected Military Platforms (1990–2000)

PLATFORM	COMPANIES[1] (1990)	COMPANIES[1] (2000)
Fixed-wing Aircraft	8	3
Launch Vehicles	6	3
Rotorcraft	4	3
Satellites	8	6
Strategic Missiles	3	2
Submarines	2	2
Surface Ships	8	3
Tactical Missiles	13	3
Tactical Wheeled Vehicles	6	3
Tracked Combat Vehicles	3	2

[1] Companies producing platforms in stated year. Not all companies produce all classes of platforms within a given platform area.

Source: Department of Defense, Annual Industrial Capabilities Report to Congress, *January 2001. Available at http://www.acq.osd.mil/ia/congress_reports.html.*

TABLE 2
U.S. Contractor Presence in Selected Military Product Areas (1990–2000)

PRODUCT AREA	COMPANIES[1] (1990)	COMPANIES[1] (2000)
Ammunition[2]	9	9
Electronic Warfare	21	8
Radar	9	6
Undersea Warfare	15	5
Solid Rocket Motors	5	5
Torpedoes	3	2

[1] Companies producing products in stated year. Not all companies produce all classes of products within a given product area.

[2] The number of ammunition companies reflects active government-owned assembly and explosive production facilities. DoD is considering reducing the number of these facilities.

Source: Department of Defense, Annual Industrial Capabilities Report to Congress, *January 2001. Available at http://www.acq.osd.mil/ia/congress_reports.html.*

While consolidation has led to the demise of brand names in the defense field, it has not led to the closing of weapon-system production lines,[69] at least not to the extent forecast by some commentators in the early 1990s. Generally, production capacity remains higher than warranted by existing contracts and projected sales. In effect, the American taxpayer is paying for more industrial infrastructure than is necessary. The excess capacity persists for many reasons, but the most significant is that defense firms can use congressional pressure to maintain low rates of production or to sell goods not necessarily requested by DoD—earning reliable profits as a politically savvy regulated public utility does.

Even after the 1990s consolidation, the largest defense firms are able to maintain multiple "centers of excellence," allowing them to bid on a wide range of platforms and integration programs. In most cases, mergers and acquisitions have broadened the new, larger defense conglomerates' portfolios of programs, but each of the once separately owned facilities continues to nurture its own core competencies. Postconsolidation integration and restructuring at the level of design teams and production facilities is loose at best. At the same time, by adding military businesses and spinning off commercially oriented facilities, the parent companies in the defense industry have typically become even more dependent on military customers than the largest defense firms were in the past.

Consolidation, even if incomplete from an economic perspective, might still have serious implications for military transformation. Many policy makers believe that less competition among defense contractors will lead to increased prices, decreased responsiveness to the needs of the military, and less innovation. This logic largely tracks with standard economic theory, but it must be applied to the defense sector with care.

Even in acquisition programs in which multiple suppliers bid for a development or production contract, political and bureaucratic forces often ensure that competition is stunted. Weapon system competitions are often not "winner take all" affairs but rather design competitions in which different firms compete only for the selection of their respective approaches. A prime is selected, but the "losers" share in profitable production. In some cases sharing means that each firm builds entire platforms or systems (as with DDG 51 destroyers); in others it means that losing firms become subcontractors to the winning firm or team of firms.[70] Politicians and industrial-base advocates often justify such production sharing by arguing that it helps to maintain firms with core defense-production capacities so that they might bid on future projects. In reality, shared production results also from the concerns of DoD and Congress about the domestic political impact of closing defense plants—often with little regard for the economic cost. The result is that the salutary effect of competition on prices touted by economic theorists is considerably diluted in the defense industry.

A related criticism of defense industry consolidation—that it may limit the industry's propensity to innovate—is tied directly to the implementation of transformation. When firms invest in innovation, their goal is to create new products and thus potential new sources of revenue. However, firms are especially interested in products that are already programmed into the defense budget; because of the up-front investment required for innovation, defense suppliers are biased toward extending the production of current systems rather than pushing the technological envelope for new products. Many critics of consolidation presume that the key motivation to innovate in the defense sector comes from industry competition—that it is firms not currently selling "legacy" systems that will be most motivated to develop new products, in the hope of replacing established sellers.

Incentives for innovation in the defense market actually differ somewhat from this traditional economic view, because the military market is a near monopsony, and the military customer demands unique products. Even in sectors in which suppliers face demand from perfectly competitive consumers, the economics literature does not provide a clear picture of the role of competition in promoting innovation.[71] Competition may provide firms with an incentive to innovate, but it reduces their capability to earn returns that recoup up-front investment; firms in competitive industries may accordingly invest less in research and development (R&D). In the defense industry, however, a powerful, single customer directly pays for the initial research and development investment and sets the agenda for innovation. True consolidation of production lines in the defense industry may even free resources that the military could use to support additional R&D.[72]

Defense industrial consolidation has been more a Wall Street financial phenomenon than a Main Street production phenomenon. As such, it will be neither a catalyst of, nor impediment to, defense industrial support for transformation. Disagreement about the advantages and disadvantages of continued defense industrial mergers and acquisitions will continue, but consolidation will not have a significant impact on industry's role in transformation.

Globalization

Despite the hype,[73] defense industrial globalization is more mirage than reality. There are three dimensions of economic globalization: trade, investment, and technology diffusion. On all three counts, there is reason to doubt that the defense sector will follow other sectors, such as the automobile industry or machine tools, much less service industries like banking and transportation, down the road toward globalization. Moreover, even if the defense industry does globalize, there is little reason to believe that globalization will either facilitate or inhibit military transformation.

There are serious impediments to higher levels of cross-border defense-related trade, investment, and technology flows. First, impediments to defense exports, from limited demand to concerns about regional instability and proliferation, are legitimate, however much the defense industry would like a freer hand to peddle its wares overseas. Second, cross-border defense industry investments, with some significant exceptions, often generate security concerns in host-nation governments, including the United States. Even if the worldwide trend toward reducing regulation and privatizing public services continues, most countries will still believe that controlling basic weapons-production facilities is prudent. Third, advanced military technologies in the United States and elsewhere are largely the product of public investment; few government officials want to share the public patrimony even with close allies—much less with countries that qualify merely as potential allies or "friends." These limits also apply to firms that produce dual-use rather than military-unique technologies, as revealed in the imbroglio over the sale of an American firm, Silicon Valley Group, Inc., to a Dutch firm, ASM Lithography Holding NV. As news accounts reported, the United States was "concerned that SVG's lithography technology—used to make lenses for spy satellites and other high-tech equipment—will be shared by the Dutch firm with potentially hostile countries such as China."[74]

In addition, defense industrial "globalization" is an uneven process. For much of the world, it consists largely of imports and limited licensing agreements to assemble, and perhaps produce, lower-end systems and components; there is no requirement for technology-intensive, transformed forces. In many cases, the potential for globalization is also constrained by the limited resources available for defense.

The most significant arena of defense industrial globalization (or, more precisely, "regionalization") lies within the North Atlantic community. Officially, NATO allies remain committed to meeting interoperability problems and equipment shortfalls by means of a strategy centered on the Defense Capabilities Initiative (DCI). Unofficially, and at the level of domestic and regional politics, the NATO commitment is less clear. Most countries want to secure a share of the overall procurement and R&D budgets for their national industrial "champions." As a result, even intra-alliance globalization remains limited by traditional political economy concerns and by the low level of European procurement and R&D spending (low, at least, in comparison to U.S. spending).

The European Union has sought to rationalize procurement strategies by allowing for the consolidation of national champions into supranational regional champions. Thus EADS, BAE Systems, Thales, and Finmecanica have emerged as the big four producers of defense equipment in Europe. For the most part, each of these firms is multinational—their research, development, and production facilities are spread across several European countries and, to a lesser extent, non-European countries, such as the United States. The four firms are increasingly entangled in a complex web of partnerships, licensing agreements, joint ventures, and other forms of collaboration. According to Mattias Axelson, EADS, BAE Systems, and Thales have "the sales and breadth of capabilities that are comparable to the leading US defence companies and each is based on a complex network of cross-border ownership structures and joint ventures."[75]

But these European firms and their joint ventures are still bound by agreements to allocate production according to national governments' levels of investment in projects, severely constraining any changes in business practices or economic efficiencies for which globalization advocates might hope.[76] Ultimately, the combination of political incentives to protect local markets, concerns about the international spread of classified information, and intra-alliance tensions over grand strategy keep European firms' operations in the United States—the aspect of globalization that would be most relevant to implementing military transformation in the United States—are almost fully independent of their parent companies' worldwide businesses.[77] Facilities located in the United States, whether owned by Americans or foreign shareholders, are managed for the benefit of the American market, and they will contribute to military transformation according to their core competencies and to the demand that doctrinal innovation sets for the products that they are good at making. Superficial defense industrial globalization will not affect these underlying realities.

Commercial-Military Integration

Throughout the 1990s, political leaders and defense industry analysts called for replacement of a defense industrial base separated from commercial industry with a single, integrated industrial base that would serve multiple customers.[78] Some of them argued that the integrated industrial base would be necessary to give defense customers access to more advanced technology under continuous development for commercial applications.[79] Many transformation advocates now argue that a military intent on transforming itself should turn away from traditional suppliers and toward firms at the forefront of the "New Economy." Others suggest that the transition to commercial-military integration has already taken place.[80] That assessment is premature; if anything, many defense firms have shed commercial divisions and product lines while acquiring more defense-related capabilities through mergers and acquisitions. Moreover, commercial firms are uninterested in commercial-military integration and thus will not do a good job of serving the customized defense market for high-end networks and nodes. Commercial-military integration may have some impact on inexpensive, low-end, simplified acquisition threshold products and on subcomponent purchases, but for the primary systems under consideration with respect to military transformation, the military customer need not and should not rely on commercial-military integration.

Some links between the commercial world and the defense industry have been developed as a result of DoD's push to integrate commercial-off-the-shelf technologies (COTS) into its defense systems as a way to reduce costs, increase capabilities, and shorten weapons-acquisition and development cycles. Incorporating those subsystems into military products can help the military avoid technological obsolescence in the face of nimble overseas competitors, who might be able to "cherry-pick" the best and most affordable commercial systems for their own limited defense investments. The defense acquisition community needs to develop the organizational capability to scan commercial innovation so that it can choose suitable technologies to integrate into weapons systems. Practically speaking, that scanning function is one of the services that DoD can and should purchase from technical advisors, systems integrators, and prime contractors. Direct contact between the military customer and commercial suppliers is not necessarily required.[81]

Fortunately, the defense industry is likely to be in a position to play this brokering role, following a trajectory of sustaining innovation in defense information technology. Since early in the Cold War, the defense industry has sought to develop high-bandwidth, secure, jam-resistant communications that combine with sensitive, multispectrum sensors to aid in rapid decision making based on incomplete data under high-stress conditions. Those performance metrics were the hallmarks of the air defense and antisubmarine warfare missions of the 1950s, and they are likewise the hallmarks of the future network of networks—at a more sophisticated level of technology.

Commercial information-technology firms that are ready to serve as component suppliers are unlikely to do anything to disrupt that defense industry role. The process of civil-military integration has not thus far progressed much beyond strategic teaming arrangements, licensing agreements, and the purchase of COTS subsystems, and the reasons for limited commercial-military integration are unlikely to change. For example, Microsoft has established a small organization for selling software to military customers and has begun to enter project teams in military development competitions—including the teams supplying network infrastructure to the CVN 77 and to the DD(X). Microsoft's role, however, is limited. First, in the DD-21 and subsequent DD(X) competition, Microsoft's main job was to provide off-the-shelf versions of its Windows NT operating system. Reportedly, Microsoft has shown almost no interest in creating specialty products to meet the needs of either its military customer or its DD-21 partners. Second, and consistent with the first observation, Microsoft's presence on the DD-21 team was largely "virtual." During the project, Microsoft apparently devoted only one full-time staff person to the DD-21 project—hardly the approach a firm would take if it were interested in learning or taking over the defense contracting business. Compared to Microsoft's overwhelming volume of profitable sales to myriad commercial customers, defense acquisition simply does not offer enough potential revenue to command much management and software engineering attention.

Many other practical difficulties inhibit commercial-military integration as well:

- Government contracting requires specialized competencies that are not usually found in the commercial IT sector (for example, dealing with Federal Acquisition Regulations, or FAR).
- Defense contractors' organizational cultures and personnel are well suited to keeping the DoD customer happy, while the more informal ways of the IT sector often produces culture shock in the staid, button-down world of DoD.
- The necessary concern of the military with secrecy, accuracy, and information assurance—more important than ever in the post–September 11 government-contracting environment—runs contrary to the instincts of many IT firms.

Recent acquisition reform efforts may make it easier for nontraditional defense suppliers to enter the defense procurement marketplace; time, experience, and the generational shifts that all organizations will encounter in the coming years will help overcome the informal barriers to cooperation between the commercial IT world and the DoD. But the incentives to surmount the barriers will remain weak, because the entire defense budget for S&T, R&D, and procurement represents a relatively small prize for American industry. As a result, defense firms will continue to guard their core competencies at the level of systems contracting, and commercial IT firms are not likely to alter their business practices to try to become systems suppliers.

Military transformation begins at the level of a system of systems, and has powerful follow-on implications for high-level systems development and procurement. Because commercial-military integration is an issue primarily at lower levels of acquisition, it need not be a major concern in our examination of the defense industrial implications of transformation.

CHAPTER FOUR

Defense Industrial Sectors

Proponents of network-centric warfare conceptually divide future military capabilities into *nodes* and *networks*. Nodes essentially correspond to what have traditionally been referred to as "platforms"—ships, aircraft, submarines, satellites, and land vehicles of various sorts. Networks refer broadly to the various ways in which platforms connect with one another to share data and information. Shipbuilding, the first of the three sectors of the defense industry on which we focus, produces Navy-unique nodes. Unmanned vehicles (UVs), our second sector case study, serve both as nodes and network components in network-centric warfare. On one hand, they serve as platforms—unmanned combat aerial vehicles (UCAVs) will carry weapons as traditional strike platforms do, and ISR UVs will carry advanced sensor payloads. On the other hand, concepts for network-centric operations envision the future employment of UVs as a means of relaying data/information to far-flung nodes. Finally, in our third case study, we examine the range of organizations that can provide systems integration services for network-centric warfare. Designing the complex technical architecture for network-centric warfare's system of systems—notably including the up-front systems engineering required, for example, to optimize use of network bandwidth and translate doctrinal rules for self-synchronization into technical requirements for data sharing—poses formidable challenges for the acquisition bureaucracy and for the defense industry.

As noted previously, our exploration of the defense industrial implications of naval transformation in these three sectors is intended to generate insights into the defense industrial implications of military transformation more generally. The possibility for innovation in these three critical sectors is a test of industry's capability to support the development of the Navy, and military, after next.

Each case study has four parts. We first describe the current industrial landscape for each sector, listing established performance metrics. We then discuss key performance metrics required to implement network-centric warfare, as well as the specific relationships between firms in the sector and the military customer. Each case study concludes with an evaluation of the types of firms needed to implement the network-centric-warfare vision.

Shipbuilding

The champions of network-centric warfare seek not only to ensure that the military after next is fully networked but also to change the types of platforms (nodes) it will operate. For the U.S. Navy the primary nodes are ships, although it obviously operates other types of platforms, including aircraft, unattended sensors (e.g., the Sound Surveillance System, or SOSUS), and unmanned aerial vehicles.[82] If the implementation of network-centric warfare requires the acquisition of nodes with performance metrics that differ substantially from those used for existing ships—that is, if the Navy demands disruptive innovation from its platform suppliers—the industrial landscape of the shipbuilding industry may change substantially, as will the Navy communities most closely linked to ships (the surface and subsurface warfare communities and the Naval Sea Systems Command, or NAVSEA). However, many of the innovations that are proposed for ships are sustaining rather than disruptive; ignoring the value of the customer relationship between established shipyards and the Navy might unnecessarily inhibit transformation. While the shipbuilding sector is likely to be substantially changed by military transformation, with new players entering the competitive mix, our analysis concludes that the current "Big Six" shipyards—Avondale, Bath Ironworks, Electric Boat, Ingalls, NASSCO, and Newport News Shipyards—have crucial competencies for transformation themselves. Transformation advocates should not be quick to abandon the skills and capabilities built up in the past.

Many NCW advocates expect a major shakeup of both NAVSEA—the Navy systems command responsible for acquiring ships—and the shipbuilding industry. They foresee Navy acquisition from outside the traditional defense industry, including domestic yards other than the Big Six, and international yards. They also foresee letting prime contracts for ships to "systems integrators," notably leading aerospace and electronics firms, rather than shipyards.

Network-centric warfare proponents argue that the Navy needs to purchase larger numbers of smaller, faster, stealthier, more lightly manned ships—in short, ships that look and perform differently from those in today's Navy. They believe that larger numbers of such ships promise both tactical and strategic benefits. Moreover, given the continued strategic requirement for expeditionary forces, next-generation warships must be able to operate close to shore in the littoral against regional adversaries practicing access-denial strategies. Future navy ships are also defined by what they will not be—they will not be the large, expensive, multipurpose, multimission ships that the

United States has built historically. Next-generation warships may well be modular. Ship designs should allow the Navy to deploy different mission packages on the same basic platforms, from antisubmarine warfare (ASW) suites to deep-strike configurations, depending on particular mission requirements. Of the major components of the DD(X) family of ships—destroyers, cruisers, and the Littoral Combat Ship (LCS)[83]—it is the latter that appears to fit most closely with the vision of NCW transformation. As Vice Admiral Mullen has said,

> They are less expensive so you can put these out in numbers and they are modular [in their mission systems]. In one area I could load up the ASW module on a handful of these and really go and attack that problem along with the rest of the architecture. If I have a mine problem, it's the same thing. So that will be a major mover for us in terms of not just getting into the ring, but staying in the ring. They [LCSs] have got to be fast, lethal, stealthy, and they have to be there in numbers. . . . LCS is not defined by size yet[,] . . . but it needs to be able to pack some punch and it needs to be able to stay.[84]

Vice Admiral Mullen's characterization of LCS requirements illustrates the performance metrics that NCW advocates hope to apply to future ship acquisition.[85] These analysts question whether the established shipyards are ready to push forward to meet the technical requirements of LCS and other network-centric platforms.

Our analysis in this section draws on Christensen's discussion of sustaining and disruptive innovation and customer-supplier relationships to examine the hypothesis that military transformation will require transformation of the shipbuilding sector as well. We consider the performance metrics associated with NCW-inspired ships and the relationship between the Navy and the shipbuilding sector. The performance characteristics of NCW platforms may require some disruptive innovations along with some sustaining ones; consequently, the industrial landscape of the shipbuilding sector may well be the part of the defense industrial base that is most changed by military transformation. On the other hand, the need for a close, familiar relationship between buyers with professional military expertise and sellers with technological expertise is likely to preserve important platform-integration business for the established Big Six military-oriented shipyards. We conclude that while the adoption of NCW principles may allow different firms to compete with traditional naval shipbuilders, the established firms will remain vital to the success of plans for building the Navy after Next.

The Shipbuilding Sector Today and Tomorrow

At first glance, shipbuilding, one of the oldest industries in the world, is a prime example of an "old economy" industry that has been, or is being, eclipsed in the postindustrial, information age. Initial impressions are often off the mark. Shipbuilding may well be an example of the emerging "new old economy," where traditional extractive and metal-bending industries are being transformed and reinvigorated by the information economy.[86] With the introduction of new design and production possibilities, old-economy industrial sectors outside the defense industry have begun to offer broad arrays of near-custom products manufactured using techniques that spread fixed costs more widely and hence reduce consumer prices. Distribution networks are also improving.

The impact of "new old economy" dynamics may be more limited in the U.S. shipbuilding industry, however, than in other old-economy industrial sectors. Naval shipbuilders are constrained by a number of factors from making the technological investments necessary to benefit from the new-old economy dynamic. Given the nature of their relationship with their primary customer, the U.S. Navy, naval shipbuilders often have little incentive to invest in cutting-edge R&D and production technologies. Customer-funded investments are typically closely tied to "stovepiped" program offices whose accounting rules make it difficult to share process improvement investments across products. Profit margins are low, especially in comparison with other industries. As a result, shareholders are relatively intolerant of infrastructure and manufacturing process investments. With the Navy buying fewer and fewer ships, shipbuilders have little hope of realizing returns on up-front technological investments during long, high-volume production runs. In one shipyard, for example, a robotic welder that was purchased as part of a move to more flexible, automated production is almost never used; the cost of programming the machine for specific parts has proven prohibitive, because it would be used only on "onesies and twosies."[87] The combination of new-old economy dynamics and transformation requirements, however, may yet transform the naval shipbuilding industry. Expansion of the fleet to include more and smaller ships as envisioned by the advocates of network-centric warfare would help justify the information technology investments that would enable shipbuilders to capture the advantages of flexible design and manufacturing.

The landscape of the naval shipbuilding sector reflects that of the broader defense industrial landscape. This sector too has experienced considerable consolidation since the end of the cold war. Until 1995, the Big Six shipyards—Avondale, Bath Ironworks, Electric Boat, Ingalls, NASSCO, and Newport News Shipyards—were owned by six different firms. With the acquisition of the Newport News Shipyard by Northrop Grumman in 2001, the six yards are now owned by a grand total of just two firms. Ingalls and Avondale had already, by 2001, become part of Northrop Grumman as a result of its acquisition of Litton. General Dynamics (GD) owns Electric Boat, Bath, and NASSCO.[88] The major American shipyards are listed in table 3.

TABLE 3
Major U.S. Private Shipbuilding Facilities—2001

Alabama Shipyard, Inc.	Intermarine Savannah
AMFELS, Inc.	Kvaerner Philadelphia Shipyard, Inc.
Atlantic Dry Dock Corporation	Marinette Marine Corporation
Bath Iron Works Corporation*	Metro Machine of Pennsylvania
Baltimore Marine Industries, Inc.	Newpark Shipbuilding
Bay Shipbuilding Company	Newport News Shipbuilding*
Bender Shipbuilding & Repair Company	National Steel & Shipbuilding Company*
Electric Boat Corporation*	Northrop Grumman Ship Systems, Avondale Operations*
Fraser Shipyards, Inc.	Northrop Grumman Ship Systems, Ingalls Operations*
Friede Goldman Offshore, East	Portland Ship Yard
Gunderson, Inc.	Tampa Bay Shipbuilding and Repair
Halter Moss Point	Todd Pacific Shipyards Corporation
Halter Pascagoula	United Marine Port Arthur Shipyard
	*The "Big Six"

As defined by MARAD. Includes both active shipbuilding yards and shipyards with build positions. Derived from data provided in Maritime Administration, Report on Survey of U.S. Shipbuilding and Repair Facilities—2001 (Washington, DC: Maritime Administration, U.S. Department of Transportation, December 2001), pp. 23–28.

Shipbuilders have been even less likely to close production lines than other defense firms. Instead, facilities have been downsized, workforces have been reduced, and production schedules have been stretched out to keep yards open and operating even during the lean times. As a result, there is significant overcapacity in the naval shipbuilding industry.

Despite the propensity to keep shipyards open, the declining number of military ships built each year and the paucity of commercial work has resulted in the precipitous decline of the naval shipbuilding industry. Each of the Big Six shipyards is underutilized. Second-tier shipyards, whether building for the Navy or for the commercial sector, generally are equally unhealthy—and even less competitive. Most American shipyards not involved in naval work are not internationally competitive and rely heavily on commercial orders that would not exist without the protectionist Jones Act, which mandates that U.S. coastal trade be carried in American-built ships.[89] High labor costs, the need for recapitalization, financial market indifference, and heavy subsidies to overseas competitors by their governments plague the industry. This weakness makes it difficult to imagine that shipyards outside the Big Six will enter the naval market in response to transformation.

The prospects for innovation in the shipbuilding industry, however, are not necessarily as bleak as they might appear. If the Navy clearly signals that it values innovation, firms will work hard to develop the most innovative ships possible. They can be expected to search the commercial world for new concepts, technologies, and materials to satisfy their customer, and to use their in-house resources to push technological boundaries. They will innovate with an eye toward the approaches taken by their competitors, who are themselves seeking to please the customer with their own strategies.

Yet left to its own devices, the shipbuilding industry is more likely to embrace sustaining rather than disruptive innovation. General Dynamics and Northrop Grumman are already encouraging the Navy to invest in incremental changes to existing designs rather than "clean sheet" redesigns. Electric Boat's proposals for a next-generation attack submarine are clearly modifications of the current *Virginia* (SSN 774) class. Newport News Shipbuilding did not resist the U.S. Navy's decision to abandon the clean-sheet approach to what was first CVX and then CVNX.[90] Evolutionary improvements in the performance of familiar products reinforce barriers to entry and allow established firms to entrench their technological advantages.

NCW and Shipbuilding: New Performance Metrics?

The most contentious part of the debate about network-centric warfare has concerned its implications for the types of ships that Congress should buy, the Navy should plan for, and the shipbuilding industry should build. At their most extreme, transformation advocates argue that traditional major combatants—from big-deck, nuclear-powered aircraft carriers to extremely capable, multirole *Arleigh Burke*–class destroyers—will not have a place in the Navy after Next. Of course, their position must be tempered by the reality that the Navy will not immediately replace all legacy ships with new ones; even a high rate of peacetime procurement would buy only a few ships per yard per year.[91] Serious current proposals plan first to demonstrate the characteristics of a network-centric force using a relatively small portion of the total fleet.

The bitterest arguments today concern the statements of requirements that will define the new ship designs—the performance metrics by which competing proposals from the shipyards will be evaluated. If the requirements that carry the day are enhancements of traditional performance metrics—to be executed by new platforms—traditional shipyards will be well positioned to develop the Navy after Next. If instead the new design requirements use new performance metrics, the change in customer requirements will likely require the acquisition community to find new suppliers.

As even a cursory review of its current fleet reveals, the U.S. Navy has long preferred large, multimission, complex, and consequently expensive, naval platforms. Military leaders naturally want to overawe all actual or potential adversaries with the most capable ships that can be designed. At the same time, political incentives have pushed the Navy toward smaller numbers of larger, more capable (and even more expensive) ships rather than larger numbers of smaller, less-capable (and less expensive) ones.[92] When faced with high cost estimates for new platforms—estimates reflecting real technological uncertainty that might undermine political support for acquisition programs—advocates naturally promise that their favored innovations can help with additional missions. That response to political uncertainty yields a kind of mission or capabilities "creep" that in turn produces complex, high-performance, multirole platforms.[93]

The Big Six shipyards have convincingly demonstrated their ability to build those high-end ships. Indeed, that is why they are the "Big Six." Their capabilities are unsurpassed. Multirole ships require the complex integration of subsystems within relatively large hulls, requiring the shipyards to develop particular core competencies. For example, the hulls of the *Arleigh Burke* destroyers are the size of those of traditional cruisers. Individual ships of that class are intended to fight antisubmarine and antiair warfare battles at the same time as they prepare for (and perhaps execute) land attack/strike missions. The result is that the design bristles with antennas, squeezes an enormous amount of equipment into a confined space, and relies on weapon systems (like vertical launch tubes) that can handle many types of missiles. The core competencies in naval architecture and complex craftsmanship that make the *Arleigh Burke*–class ships tremendously capable are evident as well in the construction of aircraft carriers, amphibious ships, attack submarines, and even combat-support ships.

Advocates of network-centric warfare emphasize a number of features of future platforms that they argue are substantially different from those of the legacy force. Some of the performance metrics for evaluating competing designs of STREETFIGHTERs, (the notional small combatants favored initially by Vice Admiral Cebrowski), the LCS, and other possible future ships are actually traditional ones—meaning that the designs will require sustaining rather than disruptive innovations. Other transformation objectives, however, establish new performance metrics, and some of the resulting ships will certainly perform less well than legacy ships in terms of traditional standards. As a result, the network-centric Navy may require disruptive innovation in the shipbuilding sector and thus the establishment of some new industrial arrangements.

Speed. Transformation advocates emphasize speed. Increased speed is supposed to be achieved through, among other things, the development of new propulsion systems and the introduction of new hull forms.[94] Yet speed per se does not represent a new goal for shipbuilders. Throughout much of naval history, speed has been at a premium. Speed has always helped warships make passages more rapidly, outrun more powerful pursuers, get within engagement range of evasive targets, and outmaneuver adversaries.

In recent decades, with the advent of missiles and the increased power of naval aviation, speed became less important than when ships exchanged gun salvos. That may change in the future, due to the increased importance envisioned for speed in the traditional matrix of trade-offs between speed and payload. In NCW, increased speed may help warships to "swarm," and intratheater transports to reach the battlespace more quickly, from over the horizon. By implication, NCW proponents may be willing to tolerate reduced weapon payloads, because, for example, strike weapons are now more lethal and more accurate. Alternately, if ground forces are less heavily equipped because their lethality arises from their connectivity to air, sea, and space-based assets—including large numbers and different types of weapons—intratheater transports might reasonably sacrifice lift capacity for speed. Note, however, that in neither of these examples is the metric of speed different from the metric used in previous periods; rather it is the use to which speed is put that is different.

Investment decision makers at traditional military-oriented shipyards will understand how to evaluate technological proposals that promise to yield faster ships. Customer demand for more speed calls for sustaining rather than disruptive innovation.

Stealth. Transformation advocates often discuss the availability of new technologies that promise to reduce the sensor signature of American platforms, including the use of composites to decrease shipboard emissions.[95] Information dominance requires improved sensors that will reveal enemy positions, but it also requires that friendly forces remain hidden from enemy sensors.

Again, however, stealth is a well-established performance metric for existing naval shipyards. Since the introduction of long-range antiship missiles that could threaten ships, low signatures (in addition to improved electronic countermeasures) have been crucial for preventing enemy target acquisition and increasing the difficulty of terminal guidance for enemy weapons. Submariners have long emphasized their advantage as "the silent service." In sum, the difference in the emphasis on stealth by today's fleet and by the next generation of warships is largely a matter of degree. Improved stealth will be the result of sustaining rather than disruptive innovation.

Engagement Range. Network-centric warfare advocates stress that the Navy must be able to meet future American strategic requirements for deep attacks against targets in access-constrained environments. In this view, naval forces must be able to mount effective attacks even when land bases (at less than prohibitive distances) are unavailable or when adversaries' attacks on fixed bases raise unacceptably the cost of close-in operations from them. Naval forces will enable follow-on forces or even halt adversary operations directly, while standing off from hostile forces.

Over-the-horizon targeting became an important naval mission with the first carrier air strikes, but it became particularly important with the advent in the 1970s of long-range antiship cruise missiles and the need for stand-off defense of battle groups from such weapons.[96] The Navy has long depended on communication and fusion of data from independent sensors and on weapons' internal terminal-guidance systems. The precision strikes from the sea against land-based targets for which network-centric warfare advocates call depend even more on the integration into fleet doctrine and equipment of new sources of targeting data and of weapons with improved terminal guidance. The performance metric for the products that they want to buy, however, is one that has existed for some thirty years.

The primary constraint on land attack from the sea has been the volume of long-range fires available. The transformation to effects-based operations and to one-shot/one-kill capabilities based on improvements in weapon accuracy, sensor resolution, and battle management speed may improve naval strike by reducing the dependence on massed fires. To this end, the DD(X) program promises (as did DD-21) new guns with longer ranges, supplemented by extended-range guided munitions, that have increased the capability to sustain precise fires.

Improved deep-strike capabilities based on new types of guns and missile systems (and their associated ISR and targeting systems) are unlikely to require new performance metrics; they simply sustain and improve existing competencies.

Battle Group Cooperation. In network-centric operations, ships will be deployed in relatively large numbers; "swarming" and "self-synchronization" based on shared access to data from sensors (both those organic to the Navy and those controlled by other services and agencies) will make operational coordination an emergent property of decentralized decision making by individual ship commanding officers. The LCS, for example, is intended to operate this way.[97]

Buying ships with the ability to operate with other ships in the battle group—especially in relatively close proximity—has been an important acquisition criterion for many years. With the development of the Cooperative Engagement Capability (CEC), efforts to improve awareness of incoming air tracks and improve cuing of the battle group's responding fires led to a major investment in high-speed, intership networking equipment. The idea that ships should fight together to maximize their effectiveness is well established.

Requirements for basic communications interoperability have forced platform designers and operators to cooperate with external groups—designers of other platforms that will serve in the same battle groups. Unfortunately, organizational boundaries have been a problem; interoperability requirements are often among the first to be sacrificed during the development process, and operators are forever asking fleet-support engineers and technicians for "quick fixes" before platforms go to sea together. If the organizational problems can be solved, calls for a common operational picture will be simply continuations of long-term demands for reducing the fog of war and improving interoperability. New and improved data-sharing may reduce the dependence on active command and control (a development that could completely reshape the Navy's operations and command structure), but new equipment to make that change possible will advance along a well-known path—it will be a sustaining innovation. However, the increased emphasis that battle group cooperation receives in connection with transformation will require adjustment on the part of shipbuilders and firms providing shipboard subsystems.

Affordability. In the face of continued budgetary constraints, the requirement for larger numbers of ships dictates that new platforms be less expensive than legacy designs. Even with the defense budget increases following September 11, the naval shipbuilding procurement account is unlikely to grow enough in the coming years to buy dozens of ships at current prices; even relatively simple warships currently cost more than half a billion dollars.

Acquisition-reform advocates have routinely tried, and routinely failed, to make cost an important performance metric for the defense industry. Buyers naturally prefer lower prices for any given capability; that was true even during the Cold War, when the pressing threat from the Soviet Union drove military requirements. Nonetheless, the acquisition community weighted combat performance higher than low cost in trade-off studies—for good reason. In the post–Cold War environment, despite the introduction of "cost as an independent variable" in acquisition regulations, the buyer continues to weight nonprice performance concerns highly in acquisition decision making; "pork barrel" politics are important in the low-threat environment.[98] Network-centric warfare may add military pressure to budget pressure for making affordability an important performance metric, but political resistance will continue. Traditional shipyards may have some incentive to adapt their designs to the goal of cost reduction, but the political safety valve will limit the likelihood that affordability will force major change in the industrial landscape.

Indeed, recent reports suggest that early plans for the Littoral Combat Ship may have difficulty meeting affordability criteria. The Navy's surface warfare directorate estimates that the first LCS will be procured in fiscal year 2005 at approximately $542 million per copy—including development costs.[99] Although this figure is lower than similar estimates for the DD-21 and the DD(X) family, it still seems too large to allow the procurement of numbers sufficient for swarming or for satisfying peacetime forward-presence requirements with smaller, cheaper ships.

Low cost has not been a traditional performance metric for the Big Six shipyards, and the requirements pressures that are driving up the cost estimates for LCS may show that the Big Six's investments in other core competencies may continue to be rewarded. However, if network-centric warfare advocates truly have their way, affordability may require disruptive innovations from the shipbuilding sector. Based on past sales of frigates and corvettes to foreign navies, some non–Big Six shipbuilders claim that they can make STREETFIGHTER-like ships for around $250 million a copy.[100] If demand for swarms of ships makes affordability truly crucial for the acquisition community, then these nontraditional suppliers may have an opportunity to break into the U.S. Navy market.

Single-Purpose Ships. Network-centric warfare advocates call for single-purpose ships, in part to eliminate the problem of "tactical instability." Some analysts argue that battle groups and amphibious ready groups are tactically unstable today in that the loss of one large, multimission platform would not only severely cripple the fleet's capabilities but be prohibitively expensive in terms of lives and resources. The high cost to U.S. forces of losing a ship provides potential adversaries a weakness to exploit and a technological aim-point—cheap weapons capable of knocking out a major American combatant.

A less complex ship might be optimized for a single mission, such as antisubmarine warfare. Losses of one or more single-mission ships, while costly, would not weaken the fleet's ability to perform its myriad of other assigned tasks. Moreover, at least in theory, a single-mission ship could be optimized to perform a particular task better than a multipurpose ship that must compromise among the performance metrics associated with different missions.

Whether buying single-purpose ships would require disruptive innovation from the shipbuilding sector depends on what other changes in platform requirements are adopted simultaneously by the Navy. Over the past several decades, the Big Six have learned to handle the complicated engineering and manufacturing necessary to fit the many complex subsystems required for multirole ships into tight spaces; as the *Arleigh Burkes* show, even a large hull can be space constrained if you wedge enough equipment into it. "Tight packing" of subsystems is a performance metric associated with multipurpose ships. Even if a dedicated, single-purpose system achieves a performance advantage relative to the comparable component of a compromise-limited, multipurpose system in part by being larger (with more computing power, cooling capacity, etc.), the total mission package of a single-purpose ship will be smaller, because it requires less functionality. Consequently, successful proposals for network-centric ships might perform less well on the "tight packing" performance metric, making the Big Six shipyards less likely to offer such proposals. Nontraditional suppliers might have an advantage in design competitions for relatively large single-purpose ships (yet still smaller than legacy types), because they have not invested in a core competency in complex naval architecture.

On the other hand, Navy doctrine writers may decide that single-purpose ships should be *much* smaller—not simply for the sake of being smaller but because of the effects of size upon signature, cost, or deployment schedule, for example—than existing multirole designs, consistent with the new-economy theme of miniaturization and with the demand for stealth. If so, the ratio of mission-system size and complexity to the hull size may not change; it may even increase. That version of transformation could reinforce the value of traditional shipyards' skills—making the shift to single-purpose ships call for sustaining rather than disruptive innovation.

Taking network-centric warfare to an extreme can highlight the potential for single-purpose ships to require either sustaining or disruptive innovation. Carried to its logical conclusion, combining single-purpose ships with enhanced battle group cooperation might increase the demand for disruptive innovation in the shipbuilding sector still further.[101] Ultimately, in a networked Navy, ships need not have many on-board capabilities, because they can distribute requests for, say, air defense to other nodes in the network.[102] NCW advocates stress that decentralization of capabilities offers benefits to the fleet; a particular node may be lost, but the overall network will remain highly capable. Conversely, however, if an individual ship must engage the enemy without access to the network—whether because of battle damage, enemy jamming, equipment failure, or unexpected dispersal of friendly units—it will be less capable than a non-NCW platform. If shipbuilders, both the yards themselves and integrators charged with populating hulls with various ship systems, need to reorient to make network-only ships, the key performance metrics will obviously shift dramatically—notably toward making sure that ships are never cut off from their battle groups.[103] The established shipyards may be reluctant to propose designs using the new performance metrics, opening the way for transformation of the industrial landscape.

If, on the other hand, shipbuilders are expected to produce ships with the full panoply of capabilities to fight independently, with network-based capabilities simply overlaid, NCW will require sustaining innovation, and the Big Six yards will be likely to maintain their dominance. The latter scenario—a less extreme version of NCW—is likely to be selected on grounds both military-operational (commanders prefer maximally capable ships under all possible fighting conditions) and political (politicians are unlikely to vote for ship designs that offer anything less than maximum protection of the American flag and of sailors serving on them). That consideration limits the likely extent of disruptive innovation associated with network-centric warfare and its emphasis on single-purpose ships.

Modularity. Modularity is perhaps the most controversial performance metric suggested by Navy transformation advocates. Ideally, new NCW ships could be optimized for missions in one environment and then rapidly reconfigured for other missions in other environments.[104] In the starkest possible terms, NCW proponents argue that the Navy needs to be able to "plug and play"—to "plug in" different payloads (dependent upon specific mission requirements and battle group composition) and continue "playing."

A less ambitious version of modularity applies only to construction; single-purpose ships built for strike, antiair warfare (AAW), ASW, or reconnaissance, for example, would share a basic design with many common parts, thereby achieving economies of scale in production. Modularity, then, might be seen as a component of the new emphasis on affordability as a performance metric for shipbuilding, providing that issues such as overhead costs of supporting modularity can be worked out. While this form of modularity would surely yield scale economies and help relieve the shipbuilding sector of some of the burdens of low-rate craft production, it would also add tremendous complexity and cost in the ship-design stage. The resulting ships would gain whatever tactical stability benefits the single-purpose performance metric will provide, but they would also require exactly the kind of demanding naval architecture and construction skills that the Big Six shipyards and the leading naval design and professional service consultancies (like SYNTEK and Vail Research & Technology, from the Arsenal Ship program) have nurtured. From an industrial-landscape perspective, this form of modularity could actually *reduce* the disruptiveness of the innovations required by transformation.

Reduced Manning. Advocates of naval transformation frequently stress the need to build ships that smaller crews can operate and fight—to reduce costs, ease problems with recruitment, and put fewer lives at risk during combat. Over the life of a ship, personnel costs loom large compared to those of design and production. Moreover, as salaries have risen in an attempt to meet recruitment and retention challenges, the failure to maximize the productivity of human capital aboard ships has become more apparent. Finally, as we will see with UAVs, one of the defining characteristics of NCW nodes/platforms is that they provide greater tactical survivability and risk fewer lives.

In the shipbuilding sector, reduced manning was not a priority in the past; if anything, ship designers were pressured in the opposite direction. Warships were and are designed and produced to accommodate a built-in surplus of personnel—to operate weapons that in peacetime operations are seldom used and to provide greater damage control capability. The developmental Advanced Gun System, for the DD(X), is the first to take personnel out of the magazine. Crew allocations are also changing now, as certain functions are moved off-ship by applying advances in telecommunications and computing. These changes may flow into future overall ship designs as part of transformation—possibly introducing a new performance metric and therefore a demand for disruptive innovations.

Commercial vessels have long operated at lower manning levels than naval ships.[105] It may be that firms with more experience building commercial ships, even smaller and much less complex vessels than warships, have core competencies in areas like automated damage control and shiphandling. These skills may prove advantageous in design competitions for the LCS and future NCW-friendly ship programs, helping new entrants establish positions in the market for U.S. Navy ships.

The list of performance metrics for ships touted by advocates of network-centric warfare includes a mixture of established and new standards for evaluating designs. Speed, stealth, engagement range, modularity, and perhaps battle group cooperation and the single-purpose platform all suggest an important role for sustaining innovations in the Navy after Next. On the other hand, the emphasis on affordability, reduced manning, and most conceptions of single-purpose ships will pull demand toward disruptive innovations that may encourage some degree of restructuring of the shipbuilding sector.

Customer-Supplier Relationships

To the extent that transformation requires some disruptive innovation, the close customer-supplier relationship between the Navy and the Big Six shipyards may delay, if not undermine, the process. On the other hand, that established relationship may help promote sustaining aspects of transformation and may allow the Big Six to serve as platform integrators, brokering connections among new entrants unfamiliar with military operations and requirements, suppliers of military mission systems, and the military customer.

Until relatively recently, shipbuilders of necessity worked closely with the Navy, if for no other reason than that the Navy reserved design and engineering functions to such organizations as NAVSEA. Further, the Navy maintained its own yards to provide the bulk of the maintenance and upgrades required by the fleet. In day-to-day terms, naval officers supervised the production of ships and submarines and then worked hand in glove with private yards throughout their shakedown cruises. The Navy has been able to cede most ship-design responsibility to private firms only because it is confident that those firms well understand its core interests.

On the other hand, NCW advocates point out that many technological advances are brewing in shipyards outside the Big Six. As a result, the advocates hope, the Navy can break its ties with established suppliers so that it can gain access to the new technologies. Existing commercial shipyards, especially in other countries, are now pushing the boundaries with new hull designs, production processes, and propulsion systems that might support the requirements of the Navy after Next. The *Visby, La Fayette, Jervis Bay, WestPac Express, Triton, Skjold,* and other innovative designs come from Swedish, French, Australian, British, Norwegian, and other overseas shipyards. Those shipyards, despite their various licensing and experimentation agreements with the U.S. Navy (and the other services), do not have ties as close as those enjoyed by Electric Boat, Newport News, and the other Big Six yards.

Yet even for the more disruptive platform innovations, established defense firms are unlikely to be abandoned entirely in the pursuit of military transformation. Generally, resistance by mainstream customers prevents established commercial firms from pursuing disruptive innovations; in the case of network-centric warfare, the innovation process began with a set of ideas in the customer community. The Navy can tailor its requirements to promote rather than hinder transformation. Private customers are constrained by financial pressures that make disruptive innovations look like poor investments. The Navy chooses its preferred investment priorities as new doctrine develops, building on its core competency in determining how best to fight; accordingly, the Navy can set requirements that encourage suppliers to work on disruptive technologies. Each established defense firm, in addition to its technical skills, has developed a core competency in working with its military customers. Firms outside the defense sector, while able to offer sophisticated technical solutions that serve nontraditional performance metrics, are unfamiliar with the language in which the military describes its requirements and do not necessarily understand the operational environment in which military products will be used.

Commercial and foreign shipyards thus may lack the real advantages that a close customer relationship would bring to the transformation process. Working with the military customer for many years has given the Big Six shipyards a good understanding of naval operations. The Big Six also follow the Navy's requirements-generation process,[106] so as to respond with alacrity and focus to new requests for design proposals. Yards outside the traditional industrial base have other customers, whose demands will limit their ability to commit all of their investment resources to the desires of the Navy. The question for the future industrial landscape in the shipbuilding sector is whether transformation proponents can engineer suitable teaming arrangements to capitalize on the platform-integration skills and customer-relationship advantages of the Big Six shipyards and also on the sources of innovation (especially disruptive innovation) outside the established industrial base.

The Big Six also understand the impact of the customer's preferences on subcontractor relationships, and they maintain large databases of suitable subcontractors. In some ways, those subcontractor relationships may be a drag on the implementation of disruptive innovations at the subsystem level, but many of the new plumbing or wiring innovations that one could imagine for ships are actually sustaining innovations. They ease space or cooling constraints or lower manufacturing costs. New-entrant shipbuilders, on the other hand, might find it difficult to manage subcontractor relationships in the way that the Navy requires (with minimum efficiency losses in the face of complex acquisition regulations). They also might find it difficult to scan the overwhelming flood of technological innovations that might find places on a major new ship design for the Navy—a much more complex process than the relatively simple platform-integration tasks that are required for commercial or foreign naval vessels.

Finally, outsiders also lack the standard operating procedures that have been developed by defense firms to manage the unique oversight requirements of selling to a government customer. For the military buyer, efficiency (minimizing transaction costs) is an important goal in the contracting process, but the government also has other crucial goals that no acquisition-reform proposal can wish away—military effectiveness, accountability for the public trust, and social policies, for instance. Efficiency is not as important as it would be for a customer in private industry, and defense firms have adapted accordingly. The transformation process needs to work in harmony with the American political process, or it will risk being derailed. The established Navy shipbuilding sector has demonstrated its ability to work within that process.

Sector Evaluation

If the Navy does choose to acquire network-centric ships, it is possible that shipyards other than the Big Six might be enticed to enter or reenter the business of building Navy ships, thereby transforming the landscape of the shipbuilding sector. A small combatant such as STREETFIGHTER, or more likely the LCS, could be built at yards other than those of the Big Six. It is no secret that the American shipbuilding industry lags behind major international competitors in a number of areas, including small-ship design and manufacturing technology. Of course, the systems-integration aspect of shipbuilding that the East Coast yards in particular have chosen to emphasize in recent years would remain an advantage of the traditional producers. But smaller yards, such as Bender and Bollinger, can overcome that advantage by teaming with systems integrators. Indeed, Halter Marine[107] has done just that in its work for foreign navies.[108]

Depending on the ultimate design, Bollinger Shipyards may be a viable contender for building the Navy's Littoral Combat Ship. Bollinger has built Coast Guard vessels for several decades. With the express purposes of learning new production techniques and, particularly, improving its ability to build aluminum hulls, Bollinger recently entered into a partnership with Incat of Australia to build high-speed aluminum catamarans in Louisiana. Bollinger also supports the *Joint Venture* HSV-X1 experimentation program in which the U.S. Navy is gaining operational experience with a new ship design for noncombat transport missions.[109] When asked whether they would be interested in bidding on and building a small combatant for the U.S. Navy, Bollinger executives were unambiguous: yes, they would bid, and yes, they would build the vessels if they won the contract. When it was suggested that they might be bought out by a larger defense contractor, they were adamant about their intention and ability to remain an independent, family-owned business with the knowledge and facilities necessary to develop smaller, faster, lighter ships for the Navy after Next.[110]

In theory at least, with some accommodating changes in U.S. law, the Navy could also farm out production of all or part of its naval ships—particularly, perhaps, small combatants and high-speed theater lift vessels (or at least their hulls)—to the most technologically advanced shipyards in Europe and Asia. At present, however, political and security concerns virtually preclude this possibility—even, it appears, on a small scale; political sensitivities about "exporting" jobs and proliferation of weapons technology are too strong. Many Pentagon officials and congressional leaders already express concern about safeguarding secrets even in domestic facilities. Yet joint ventures, teaming, and licensing arrangements that would allow the U.S. government and American shipbuilders to develop cooperative relationships with foreign yards are feasible. Bender and Bollinger have reached agreements with Australia's Austal and Incat, respectively, and transformation may help them both break into the defense industrial base and also contribute to the globalization of the defense market. On the other hand, it is possible that the globalization inherent in these international joint ventures may actually constrain their ability to enter the U.S. military shipbuilding market.[111]

The Big Six could also face a challenge from systems-integration houses as the Navy moves to a network-centric future, but this challenge is considerably less likely to revolutionize the defense industrial landscape than is the prospect for entry by commercial or foreign shipyards. Many people see an intuitive connection between network-centric warfare's shift in emphasis from platforms to networks, and a shift in emphasis from hulls to internal electronics in shipbuilding.[112] Consequently, traditional prime contractors in the aerospace and electronics sectors of the defense industry hope to take the lead role in integrating naval platforms in the future. Raytheon and Lockheed have already filled this role in bids on the LPD 17, CVN 77, and DD(X) programs, and the teams of bidders for the U.S. Coast Guard's current DEEPWATER project may well be harbingers of a transformation of defense industrial relationships and the defense industrial pecking order.[113] On the other hand, disputes are already beginning concerning whether problems being experienced on such contracts as the LPD 17 are natural "teething troubles" for such new relationships or are more fundamental, indicating that in the complex shipbuilding world aerospace contractors may have involved themselves in a business that they do not truly understand.

It does seem clear that trying to force this particular change on the shipbuilding industry is undesirable, because it throws away the benefit of a core competency of the established prime shipbuilders. They actually specialize in the complex integration of electronics into naval platforms—dealing with space, power supply, cooling, antenna placement, and other issues that must be balanced with structural demands of ship design. Moreover, the leading naval shipyards have established procedures for subcontracting for naval electronics systems—sometimes even working with units of the same aerospace primes that are trying to move into the ship systems integration role. Their ability to solicit bids from suppliers of subsystems and manage subcontracts that meet defense acquisition requirements is a key comparative advantage relative to the potential for commercial and foreign shipyards to serve as prime contractors.

Unmanned Vehicles

Unmanned vehicles (UVs) are ubiquitous in joint and service visions of the military after next. A striking number of unmanned aerial, surface, and subsurface assets populate depictions of the future battlespace.[114] Unmanned vehicles are to bring a number of critical capabilities to the fight. They will be employed as ISR assets, communications relays, and precision strike platforms. Many, though not all,[115] of the tasks envisioned for UVs in the future are currently performed by manned platforms or space-based assets. In this report we focus specifically on unmanned aerial vehicles (UAVs) because (1) they represent the most highly developed segment of the general UV market, and (2) they are expected to perform the widest range of future missions. Existing UAVs such as General Atomics' Predator and Northrop Grumman's Global Hawk have played important roles in recent conflicts ranging from the Balkans to Afghanistan. If NCW becomes the organizing concept underlying military transformation, the Navy and the other services will acquire a large number of unmanned platforms that will be assigned an increasingly greater number of roles, missions, and functions.[116]

Our second sector comprises firms now designing and building, or capable of designing and building, unmanned aerial vehicles. This industrial sector has emerged only recently, unlike shipbuilding, which has existed as a distinct industrial sector for hundreds of years.[117] However, the type of firm that will supply UAVs to the military after next remains an open question. Existing firms like Northrop Grumman's Ryan Aeronautical[118] and General Atomics Aeronautical Systems, Inc., enjoy the advantage of having built deployed UAVs. Yet they may not remain the suppliers of choice in the future, since the critical performance metrics for unmanned systems are not entrenched. In theory, at least, competing firms could offer better solutions to outstanding technical challenges, thereby setting the standard for future acquisition. In this section we survey the UAV industrial landscape, identify possible performance metrics for current and future unmanned systems, examine the nature of customer-supplier relationships in the sector, and explore the future of the UAV sector in the transformation process.

The UAV Sector Today and Tomorrow

Although UAVs have been used by the U.S. military at least since the Lightning Bug was deployed in Vietnam, many subsequent efforts were canceled, including the Aquila, Amber, Medium Range, and Hunter.[119] Even such relatively successful UAVs as the Pioneer were deployed in only limited numbers and suffered from performance limitations. Why the United States has not used unmanned aerial vehicles more extensively, thus nourishing an industrial sector to develop and produce them, is unclear. As a RAND report noted, "It has been technically possible to build generic UAV platforms for several decades, and many have been built and used as aerial targets and reconnaissance drones."[120] Analysts stress institutional and cultural resistance to UAVs as well as an absence of clear demand due to competition from a diverse array of successful platforms for performing similar missions.

Yet by defense industry standards, the industrial landscape of potential UAV manufacturers is thickly populated. More than thirty firms were active in the UAV sector in 2001 (see table 4). During the past decade, most DoD and service R&D and procurement spending on UAVs has gone to Northrop Grumman's Ryan Aeronautical, Boeing, and General Atomics. Several smaller firms, such as AAI Corporation and Aero-Vironment, Inc., have built relatively successful UAV prototypes and experimental platforms for DARPA, NASA, the U.S. Army, and other government agencies. Three other types of firms may also have the expertise to enter into the UAV market in the future: (1) traditional defense firms, such as TRW, that have built UAVs in the not-so-distant past;[121] (2) start-up firms that may offer innovative solutions to long-standing technological challenges facing UAVs; and (3) foreign UAV manufacturers.

There is a thriving international UAV market. In contrast to other areas of defense acquisition, European militaries, individually and as part of NATO or the European Union's "Rapid Reaction Force," have invested considerable resources in UAVs. A number of competitors to American-built UAVs are already on the market, partially in response to this European demand. Nineteen companies in France, Germany, and the

TABLE 4
U.S. Private Sector UAV Manufacturers

COMPANIES	UAVs	COMPANIES	UAVs
AAI Corp.	Shadow	Insitu Group	Seascan
AeroVironment	Black Widow, Centurion, Pointer, Hiline, Pathfinder	Kaman Aerospace Corp.	K-Max
Advanced Hybrid Aircraft	Hornet, Wasp	Lockheed Martin	420K, LOCASS
Advanced Soaring Concepts	Apex	Meggitt Defense Systems	Sentry
Aurora Flight Sciences	Chiron, Perseus, Theseus, UCAV Demonstrator	Micro Craft Technology	LADF
BAE Systems	R4E SkyEye	Mission Technologies, Inc.	Backpack, Mini-Vanguard, Vixen, Hellfox
BAI Aerosystems	Aeros, Exdrone, Javelin, Tern	Northrop Grumman Ryan Aeronautical	ADM-160 MALD, BQM-74C, BQM-145A, Global Hawk, Fire Scout, Scarab, Sea Ferret, Star-Bird, X-47 Pegasus
Bell Helicopter Textron	Eagle Eye	Orion Aviation	Seabat
Boeing	Dragon Fly, X-36, X-45A	Raytheon Electronic Systems	AN/ALE-50
Bosch Aerospace	AURA, SASS-LITE	SAIC	Vigilante
California Unmanned Vehicles, Inc.	CUV SLURS	Sanders Defense Systems	MicroSTAR
Daedalus Research, Inc.	Dakota	Scaled Composites, Inc.	Proteus
Dragonfly Pictures, Inc.	DP4	Sikorsky Aircraft	Cypher, Dragon Warrior
Freewing Aerial Robotics, Corp.	Freewing Tilt-Body	Skysat Systems Corp.	High Altitude Airship
Frontier Systems, Inc.	Hummingbird	TCOM LP	15M, 32M, 71M
General Atomics	Altus, Gnat, Predator, Prowler	Thorpe Seeop, Corp.	RM1 Spinwing
GSE, Inc.	Vindicator	USBI, Co.	Dragon

Source: Kenneth Munson, ed., *Jane's Unmanned Aerial Vehicles and Targets, Issue Seventeen* (Coulsdon, Surrey, UK, and Alexandria, VA: Jane's Information Group Limited, December 2001), pp. 194–312.

United Kingdom alone were actively engaged in the UAV market in 2001. Israel also has a long history of building UAV systems. Its operational successes with UAVs dates back to operations over Lebanon's Bekaa Valley. Seven Israeli firms are presently active in the UAV business.[122]

NCW and UAVs

Network-centric warfare envisions employing UAVs in many roles: as long-endurance communication relays (supporting the network); as small, inexpensive, fast-moving, hard-to-detect sensors (nodes to support a common operational picture); and as platforms for delivering precision strikes against targets that are too difficult or dangerous for manned platforms to reach. Moreover, other service and joint vision documents and transformation roadmaps suggest that UAVs will play other important roles in the future as well. If anything, NCW and Navy planning documents place less emphasis on UAVs than do those of the other services—not because the Navy is less enthusiastic but because it foresees a rosy future not only for UAVs but also for UUVs and USVs. Through these diverse uses, military transformation will for the first time establish core performance metrics for the UAV industry.

Emerging UAV Performance Metrics

Which types of domestic and international firms will prosper as UAV usage in the military matures will depend largely on the evolution of those metrics. Firms with the technical capacity and experience necessary to meet emerging measures of success will, in all likelihood, win future design and production competitions—providing, of course, that the services budget sufficient resources.

Two general schools of thought on UAV performance metrics can be discerned. According to the first, UAVs have been built for years; they resemble other, already successful products (whether autopilots on commercial aircraft or various forms of cruise missiles and unmanned target drones). Thus, once the military commits to fielding UAVs and determines what roles they will play in future conflicts, they can be built in greater numbers. Performance metrics are not a significant issue; they are the same as for "similar" systems—the implication being that the technological challenges can be solved with sustaining innovation.

The second school believes that UAVs are unique, that their performance metrics will not easily transfer from other types of systems and platforms. When we asked government personnel involved in UAV acquisition programs, industry officials, and outside observers about whether there are generally accepted and well understood performance metrics for these systems, we received a variety of answers. This lack of consensus reflects real uncertainty. We have identified a set of ten possible performance metrics.

Mean Time between Failures. This metric refers not to survivability against enemy countermeasures (although this ultimately matters too) but to ability to remain in the air without experiencing either catastrophic failures or operator errors from which recovery is not possible. This metric can be applied to all military systems (and commercial systems as well). It does not distinguish UAVs from alternative platforms and systems.

Mean time between failures may, however, help us distinguish successful UAVs and their manufacturers from their less successful competitors. General Atomics executives claim that the Predator is the only UAV that has demonstrated exceptional success on this measure. They further claim that the performance of the soon-to-be operational Predator B will be even better.[123] Whether their claims are accurate is subject to interpretation; from recent reports we know that in the Afghan campaign at least twenty-five Predators have "crashed [due] to mechanical failure, weather, or operator [error]" or to enemy fire.[124] Other UAVs, both deployed and in the OT&E stages, have also been plagued by numerous failures. Global Hawks have crashed several times during that program's short life; reported causes range from quality control problems to operator errors. Certainly, if UAV manufacturers cannot meet this basic performance characteristic, they will not be viable producers for the military after next.

Affordability. Affordability is purported to be a key advantage of unmanned systems of all kinds. If NCW requires populating the future battlespace with numerous UAVs performing a diverse array of missions, cost will be an issue. The Global Hawk, for example, began with an ACTD budget goal of ten million dollars per copy, yet a recent estimate put the out-year cost at roughly seventy-five million per system.[125]

Like mean time between failure, this metric appears to apply to all military systems. However, affordability may play a special role in determining the attractiveness of an emerging technology for greater or more varied roles in future military operations. For UAV manufacturers and transformation advocates attempting to attract new users with new types of missions, price is a selling point. If UAVs perform well enough *and* stay within budget constraints, they may be more attractive for some end users than systems that perform spectacularly but remain prohibitively expensive. On the other hand, some supporters of UAV acquisition fear that exaggerated expectations of affordability are one of the reasons that they have not yet been widely adopted by military forces.[126] The UAV development process is vulnerable to the same "gold plating" pressures that plague other programs. An increase in costs is likely to reduce the number of systems that end users will be willing or able to acquire.[127] As a result, UAV advocates may choose to promote their products by minimizing the weight placed on affordability as a performance metric.

Reduced Manpower Requirements. UAVs do not reduce personnel requirements as directly as the caricatures provided in news reports often suggest; while UAVs have no flight crews, they still require remote operators (equivalent to pilots) and maintenance and support crews. They have substantial logistical tails; nevertheless, their "tails" are smaller than those of forward-deployed manned aircraft.

Fielding and sustaining a fleet of manned aircraft is incredibly labor intensive; everything from training to maintenance is required, involving thousands of hours of manpower. UAVs may at least reduce the time and money needed for training. UAV operators (with their associated support personnel) can accomplish much of their training and skill development on simulators, in contrast to the hundreds of hours in the air that manned-aircraft pilots must spend developing and maintaining their proficiency. In addition, UAV system developers can readily incorporate simulator functions into flight control stations.

Flight Endurance. The duration of manned aircraft flights is constrained by physical limitations of the flight crew, among other factors. Short of rotating aircrews already onboard, manned aircraft cannot simply linger for long periods without losing their effectiveness (regardless of the ability of the aircraft itself to remain aloft and the amount of "down time" an aircrew might enjoy during a mission). UAVs offer more flexibility. First, multiple teams of operators at various remote ground control stations (GCSs) can operate the same UAVs in succession. Duration of flights is then only limited by the power supply available to the craft, the possibility of mechanical failure, and, in combat situations, the UAV's survival. Second, recent operational deployments suggest that software improvements could reduce the demand for operator intervention and the need to transmit certain types of data during UAV flights.

Power Supply. The quality of power sources is another possible UAV performance metric. Power affects both the ability of the aircraft to fly and the types of mission packages that can be carried on the basic UAV airframe. Flight duration, cruising speed, communications capabilities, and sensor strength, for instance, all depend on the amount of power generated by the UAV's engine, the fuel efficiency of that engine, the quality and longevity of the power source, and the stability and continuity of the power flow. Many ISR packages require a great deal of energy to operate; the development of new power supplies may expand the range and quality of the ISR packages that can be bundled into UAVs.

Mission Payload. A UAV is only as effective as the mission payload it carries, whether the payload is a sensor suite, a communications system, or a weapon. For UAV manufacturers, the task is to design or purchase the best payloads (from the perspective of the end user) and integrate them onto the platform. Although engineers can make trade-offs among the various desired performance characteristics of UAVs, designs should maximize the ability to carry mission systems; UAV makers must learn to choose compact, energy-efficient subsystems.

Datalink Quality. The ability of the UAV to communicate with the GCS and with other platforms and sensors in and around the battlespace is critical. If, for example, the UAV carries an ISR payload, it needs to be able to transmit data at times and in formats useful to consumers. Relevant trade-offs include where to process the sensor data (on board the UAV, at the consumer's location, or at some intermediary point) and how often to download information (continuously, at scheduled intervals, or at critical points determined by the characteristics of the acquired data). In all cases, the accuracy and reliability of download and upload technologies and protocols must be sufficient to meet the needs of users. Datalink quality also determines the ability of the remote human operator to control the craft, especially under adverse conditions of, say, weather, terrain, or enemy action.

Plays Well with Others. Another key constraint on existing UAVs is the requirement that they fly safely in the same airspace as manned systems and other UAVs.[128] This requirement would appear relevant for all platforms, not just unmanned systems. At present, situational awareness is more limited for UAVs than it is for traditional manned aircraft,[129] but as testing and operational experience with UAVs accumulates, their capabilities should catch up—especially since manned aircraft face constraints too (due to, for example, canopy design). Not all manned aircraft can easily accommodate extra personnel dedicated to monitoring sensors; UAV ground stations can more readily be expanded to relieve the burden on the pilot, and their computer processing power can readily be augmented to enhance data management capabilities.

Mission Controlled by End User. One of the driving forces underlying demand for UAVs in recent conflicts has been the perception (and the reality) that other types of assets performing similar functions are not under the direct command and control of the end user. Deployed Army and Marine units sometimes find it difficult to task ISR and strike assets controlled by the Air Force and the Navy. UAVs directly attached to local commanders will, by definition, be more responsive. Insofar as UAVs can be designed to facilitate interactions with the ultimate consumers of their services, this metric will play a critical role in determining whether UAV projects will find sponsors.

Optimizing operational control of UAVs requires that numerous technical and organizational issues be resolved. For potential UAV suppliers, key technical issues include where to locate processing and analytical capability and how to deliver data to the end user.

Safety of Personnel. Of all the potential performance metrics that may determine outcomes for the UAV sector, the elimination of risk to pilots appears to matter most. With the important exception of space-based assets performing ISR and communications-relay missions, UAVs compete with systems that, by definition, put their operators at risk. Navy and Air Force aircraft that perform close air support or deep strikes may be shot down, risking the death or capture of the flight crews. By contrast, if a Predator equipped with Hellfire missiles is shot down, only equipment is lost. The emphasis on saving lives is especially important when mission performance depends on close proximity to the battlespace—a factor that stimulates demand for tactical UAVs.

Which performance metric, or group of performance metrics, will set the standards for UAV designers and builders will be revealed over the next several years as the results of testing, experimentation, and operational experience become available. Although most potential UAV metrics appear to require sustaining innovations by existing firms, it is possible that new firms will prove more adept at providing the Navy with products maximizing particular disruptive performance metrics. For instance, a new firm could develop UAVs capable of reliable autonomous operations. Greater autonomy could prove disruptive, because most UAV manufacturers have not, to date, made this their primary focus or invested heavily in the technologies that would allow for it. Yet successfully resolving the autonomy challenge might be attractive to consumers; it would, for example, help reduce the manpower required to operate UAVs.

Until questions about specific disruptive and sustaining performance metrics are resolved, firms with proven track records will remain in the driver's seat, while late entrants and start-ups will seek to break into the marketplace based on new technologies and skill sets adapted from the design and production of other weapons systems. The performance metrics that ultimately set the standard will determine the extent to which UAV suppliers will be expected to provide sustaining or disruptive innovations. In this still evolving sector, it is not yet clear whether customer requirements will be met predominantly by sustaining or disruptive innovation. Both are likely to be required.

Customer-Supplier Relationships

Performance metrics alone will not determine which types of firms will thrive when and if the U.S. Navy and the other services make wholesale purchases of UAVs. Existing and emerging customer relationships will also shape the future industrial landscape. Because the services have not yet purchased many UAVs, neither acquisition organizations nor their technical advisors have formed close relationships with particular contractors in this sector. Instead, several firms have modest track records, and a larger group of companies can claim either direct experience with, or demonstrable technical potential for responding to requests for, proposals. Translating these limited ties into a comfortable working relationship with military customers will help determine which types of potential UAV suppliers are most able to match their technical skills to the operational requirements of network-centric warfare.

The relationship-building experience of the contractors that are currently selling UAVs to the U.S. military shows the nascent state of the sector. General Atomics explicitly hopes to profit from its track record; the Predator and its follow-on, the Predator B, are already flying and purportedly can be adapted in short order to fulfill most UAV mission requirements if the military communicates them clearly to General Atomics. However, General Atomics executives complain sharply that the U.S. Navy does not recognize the demonstrated superiority of their products and has declined to develop doctrine and establish requirements by actually flying Predators. Instead, by the executives' account, the Navy insists on more paper studies and more experimental prototype development before buying any UVs, specifically including Predator.[130] As much as General Atomics would like to have a close, trusted relationship with the U.S. Navy, it is clear that it does not have an inside track.

Having acquired the established UAV manufacturer Ryan Aeronautical and subsequently developed the Global Hawk, Northrop Grumman appears to have bought itself credibility within the UAV community. However, Global Hawk has not yet demonstrated peak performance for extended periods, and chronic equipment failures suggest that Northrop Grumman will have to work hard to maintain the position that it has attained.[131] While Congress has mandated that the Navy purchase several Global Hawks as a means of experimenting with unmanned systems as a supplement to manned patrol aircraft, the Navy has not shown a propensity to cooperate closely with Ryan Aeronautical. Though the program was subsequently revived, the Navy terminated the Fire Scout, Ryan Aeronautical's prototype vertical-takeoff UAV, at the flight test stage, because the acquisition community lacked an experimentation plan under which to use the craft to develop future mission requirements.[132] All the equity that Northrop Grumman really has in this respect is its overall corporate commitment to position itself as "the RMA firm." It invests more effort than most other defense firms in understanding the nature of network-centric operations. If the parent company's pro-transformation stance filters down to the UAV division, it may have an advantage in responding to future requirements for the Navy after Next.

Boeing too has gone to great lengths to demonstrate network-centric expertise. It has produced extensive independent analyses of NCW and used them to supplement NWDC strategic planning documents. Boeing's future as a producer of UAVs hinges on the success of its prototype UCAV, the X-45.[133] If the X-45 turns out to be a technical success, it will help Boeing to establish the capability to meet key UAV performance metrics. But even if the X-45 itself does not perform well, the program could provide Boeing with inside information and a comfortable relationship with evolving military requirements for UCAVs.[134]

The problems that existing UAV producers face in their relations with the Navy, however, pale in comparison with the difficulties facing start-up firms and other small companies. Officials from mainstream UAV manufacturers and the Navy's UAV R&D and acquisition organizations joke about the stereotypical wild-eyed tinkers working in their garages to produce "big model airplanes," which they hope are the same thing as militarily useful UAVs. It would be tempting to dismiss their scorn as uninformed, but some small UAV ventures seem to invite such criticisms. Many UAV start-ups do work out of "garages," employ engineers with little understanding of military requirements, and use business models often hinging on joint ventures with, or acquisition by, other manufacturers, particularly firms with established relationships with the military customer. In short, it would appear that there are long odds against serious challenges by new entrants for leadership of the UAV industry.

Sector Evaluation

In all likelihood, growth in the UAV market will not require disruptive innovation, in the sense in which the term is typically used. Defense firms have a long history of producing unmanned systems—from Vietnam-era versions of contemporary UAVs to cruise missiles.[135] Boeing, General Atomics, and Northrop Grumman have already developed significant UAVs, and they collectively, at least initially, have a lead in weaponization of unmanned vehicles. Some quality metrics for such systems are well known, although many high-performance UAV technologies are still immature. But because almost none of the past programs has entered full-rate production, current defense aerospace manufacturers do not have much investment in UAV-related customer relationships. Technically skilled new entrants, however, have even less familiarity with military culture or warfighters' professional expertise. The net result is that the future defense industrial landscape in the UAV sector is wide open.

At least until very recently, the industry's comfort level in producing UAVs appeared to exceed that of the military in using them. Resistance to UAVs remains, and will remain, among defense industry stalwarts and within some Navy and other service communities, especially when UAVs threaten missions normally assigned to manned assets. Much of this resistance plays out in the planning, programming, and budgetary processes. In recent years, interested congressional leaders, such as Senator John Warner (R-Va.), have ensured the availability of additional resource for UAVs, but the Pentagon's relative disinterest in the fruits of that investment has delayed the development of performance standards. The future of this sector is highly dependent on the strength of the military services' presumed commitment to UAVs as part of military transformation; S&T and R&D monies sufficient to overcome technological hurdles and operational challenges have not been allocated.

As transformation advocates and planners envision new missions for UAVs and other unmanned vehicles, they will eventually develop a full array of performance standards for the defense industrial base. Even relatively successful current projects face significant "capabilities gaps." For example, the current generation of UAVs requires intense human operator involvement, which limits their usability. They are also highly vulnerable to enemy attack and countermeasures, and they exhibit limited fault tolerance, making them prone to crash. Other problems are certain to emerge as requirements for UCAVs work their way through the acquisition system for the first time. Firms will need to (1) develop for all types of UAVs command-and-control systems that will allow them to operate in a battlespace populated with manned systems and (2) provide future warfighters reasons for confidence that UCAVs will be able to distinguish legitimate targets from noncombatants. Partnerships among established defense contractors, start-up UAV specialists, and in-house experts in military operations can be expected to set the pace for the contribution of unmanned vehicles to military transformation.

Systems Integration

The network-centric transformation vision relies heavily on the ability of various nodes to share information in real time using a range of interconnected networks. Achieving the NCW vision will require lashing networks together, maintaining networks in the face of constant change, making intelligent trade-offs among competing system designs, and assigning operational roles to various platforms. Transformation thus places a high premium on systems-integration skills and upon the public and private organizations that possess them.

A basic definition of systems integration emphasizes interoperability—the requirement that each military system work in concert with other systems, on the basis of sufficient communication across well-defined interfaces. Network-centric warfare concepts obviously stress such intersystem compatibility; casual discussions of systems integration in the context of transformation, in fact, often refer only to interoperability requirements.[136] However, ensuring interoperability is only one part of the systems integrators' task. Systems integrators are responsible for a number of roles during the overall acquisition process, beginning with translating objectives derived from military doctrine into technical requirements suitable for launching acquisition programs. The key part of this process is making trade-offs of capabilities among various systems—given a set of desired capabilities, which component of the system of systems should perform each of them? In the current, early stages of thinking about network-centric warfare, systems integration should define the nodes that make up the network, the capabilities that will be essential for each type of node, and the number of nodes that must participate in various operations. Later in the acquisition process, systems integrators must maintain control of technical standards and interfaces (ensuring interoperability), manage cooperation among contractors and subcontractors, test products and their subcomponents, and support users' efforts to customize and modernize products as missions and technologies evolve.

There are several levels of systems integration in the defense sector; all of them involve making choices between technical alternatives and establishing links between disparate equipment so that heterogeneous parts can operate together. First, at the "lowest" level, *weapon-system integration,* various components, often supplied by subcontractors, are tied into a single product (e.g., a surface-to-air missile or a fire-control radar).[137] Certain key facilities owned by the prime defense contractors specialize in this type of systems integration (such as Raytheon in Tucson, Arizona, for missiles, or Northrop Grumman in Linthicum, Maryland, for radars). Second, *platform integration* combines various types of equipment (weapons, propulsion, sensors, communications, etc.) into mission-capable assemblages. This second process is not necessarily more or less complex than weapon-system integration, nor does it necessarily add more or less value; different types of systems integration must be analyzed on a case-by-case basis. But again, some prime contractors (Lockheed Martin Aeronautics in Fort Worth, Texas, or General Dynamics' Bath Ironworks in Bath, Maine) define platform integration as one of their core competencies.

The real emphasis in transformation—and the level of systems integration that is now most ardently pursued by defense-oriented organizations—is *system-of-systems integration,* or *architecture-systems integration.* It connects different types of platforms so as to facilitate cooperative military operations; it constitutes the technical counterpart to the military services' operational expertise (knowledge of how to fight). It essentially translates doctrinal statements of objectives into sets of requirements that can be written into the acquisition community's contracts with industry; it involves broad trade-offs among different technical approaches—for example, hardware-versus-software solutions, or a decision on whether to transmit raw or processed data across the network. Historically, system-of-systems integration has been accomplished by organizations within the military services (e.g., laboratories that support systems commands, like the Naval Surface Warfare Center, Dahlgren Division) or closely allied to them (specialty organizations, including federally funded research and development centers [FFRDCs] like the MITRE Corporation). Network-centric warfare's emphasis on simplified platforms, distributed capabilities, and interconnection of military assets via advanced communications networks will force the acquisition community to rely more than ever on first-class system-of-systems integration.

Military-oriented systems-integration skill is based on advanced, interdisciplinary technical knowledge—sufficient understanding of all of the systems and subsystems to make optimizing trade-offs. It also requires a detailed grasp of military goals and operations as well as a reservoir of trust that bridges military, economic, and political interests. Some systems-integration (SI) organizations also have some production capabilities (which may be either an advantage or a liability to the integration process), but systems integration is a separate task from platform building and from subsystem development and manufacturing.

Systems integration is an independent sector of the defense industrial base, but one with porous boundaries that sometimes allow members of other sectors (e.g., platform builders) access. Different combinations of systems-integration capabilities are found in traditional defense industry prime contractors, specialized systems integration houses, FFRDCs and other quasi-public organizations, and the military laboratories. Because all these organizational types understand the crucial role of systems integration in transformation, most are maneuvering to establish their credibility as systems integrators; for example, prime contractors justify acquisitions on the grounds that they contribute to a "systems integration capability," and military laboratories rewrite mission statements to emphasize systems integration.[138]

Organizations that can provide systems integration services should have a key, early role in implementing transformation. Objectives for projects in other sectors of the defense industry—such as for platform makers, like shipbuilders—will flow down from the overall definition of the network-centric system of systems. Early in the transformation process, systems integrators need to determine what capabilities are necessary for each type of node in the network, in light of the technical, operational, and economic implications of how capabilities are distributed. This job is one for which the massive, complex Cold War defense effort left the United States well prepared. Organizations that specialize in system-of-systems integration were established in those years as part of the ballistic missile and air defense programs, and in cooperation, they also played vital roles in developing equipment for maritime strategy, missile defense, and other system-of-systems missions. Network-centric warfare calls for sustaining innovation in the systems integration sector; transformation advocates need to recognize and exploit the established skills at the front end of the process.

The System-of-Systems Integration Sector Today and Tomorrow

Many organizations have at least some expertise that might contribute to system-of-systems integration for the Navy (for a list of examples, see table 5).

As the customer for military equipment, the Navy must define projects' objectives, but the actual technical system-of-systems integration task is very difficult for the Navy itself to accomplish. The acquisition community's core competencies, resident in the system commands, are in understanding government regulations and monitoring suppliers' compliance with cost, schedule, and other contractual terms; acquisition agents are usually not expert in state-of-the-art technologies and the innovative capabilities of various firms. The Navy's old technical bureaus were phased out during the second half of the Cold War, and technical tasks were increasingly outsourced to private industry.[139] Systems commands can still draw on expertise from subsidiary laboratories (e.g., SPAWAR Systems Center, San Diego, for C4ISR), which maintain important niche capabilities, research expertise, and key physical assets (e.g., model basins) required to develop and test new designs "end to end." Unfortunately, the relationship between science-oriented military laboratories and regulation-oriented systems commands is often tense. Scientists often feel that the continuity of their research

TABLE 5
Examples of NCW-Related System-of-Systems Integration Organizations

	GOVERNMENT	PRIVATE, NONPROFIT	PRIVATE, FOR-PROFIT
Analysis	System Commands (SPAWAR, NAVSEA, NAVAIR)	Center for Naval Analysis, Institute for Defense Analysis, RAND	ANSER, TASC, Booz Allen Hamilton
Scientific Research	Naval Research Laboratory, SPAWAR Systems Center, San Diego*	APL, Lincoln Laboratory, Software Engineering Institute	
Technical Support	SPAWAR Systems Center, San Diego*	APL, MITRE, Aerospace Corporation	SAIC, SYNTEK
Production			Lockheed Martin—Naval Electronics and Surveillance Systems, Raytheon Command Control Communications and Information Systems
Testing and Fleet Support	SPAWAR Systems Center, San Diego*		

* Each of the Navy's acquisition system commands has related technical organizations equivalent to the SPAWAR Systems Center—for example, the Naval Air Warfare Center, China Lake, and the Naval Surface Warfare Center, Dahlgren.

Note: Some organizations have additional small-scale activities that give them limited capability in other boxes in the above matrix—for example, SPAWAR Systems Center, San Diego, manufactures Link 16 antennas for surface combatants. The above designations are intended to capture organizations' core competencies rather than ancillary work.

and their technical skills are undermined by frequent "cherry-picking" of researchers out of the laboratory and into the system command itself. For their part, systems command personnel tend to believe that scientists should support their immediate needs for technical advice and technologies rather than pursue research projects that may or may not pay off.

This difficult interface between "pure" science and system acquisition is a challenge for all forms of technical advisory organization—not just for the military's in-house laboratories—but the difficulty is magnified within the military chain of command. Internal Navy technical capabilities are, on the one hand, constrained by civil service rules, which prevent the Navy from competing effectively for the services of many top scientists and engineers. On the other hand, the same rules protect internal technical staff from competitive and budgetary threats. The operational Navy often perceives the Navy laboratories and technical advisors as less cooperative than the highly responsive scientists and engineers in private defense industry, who can be induced to work hard for the military through appropriate contractual compensation. As a result, the operational Navy often fails to support the Navy laboratories aggressively.[140] This tension may be exacerbated by "industrial funding," which forces laboratories to seek "business" from within other parts of the Navy, other government agencies, and even private industry, by drumming up external contracts and participating in various project "teams," usually with specific, short-term deliverable products.

Warfighters do support the laboratory system, but only in a particular way that undermines the labs' ability to conduct analyses of alternatives and make high-level trade-offs among technical approaches. The Navy's system centers are very good at fleet support. But those close ties to quick-reaction demands of the fleet do not comport with the standardization and interface stewardship role of the systems integrator, and the skills that enable fast fixes in the field—especially of particular systems or subsystems—are not the same as the skills that produce thoughtful optimization of the system of systems.

Laboratories emphasize testing system performance, confirming that prototypes meet specifications, and determining which of several submissions best meets military acquisition criteria. This emphasis permeates these organizations so strongly that several scientists in military laboratories that we interviewed even defined systems integration precisely in terms of testing performance and interoperability. While they understand the importance of technical advice during the analysis of alternatives before the definition of performance-evaluation criteria, laboratory personnel particularly value feedback from testing physical systems in improving the ability to define later projects. On the other hand, organizations other than in-house labs do extensive testing and prototype evaluation as part of system development, even though they do not perform the final stage of customer acceptance tests. If in-house scientists are right that testing can help maintain technical skills and reveal important lines of evolutionary research, it might be desirable to sell the major testing facilities—the remnants of the unique intellectual and physical capital inside the military—to the organizations that can act as full system-of-systems integrators. The goal would be to leave the systems commands with enough technical competence to act as "smart buyers" that can react to technical advice and choose among systems integration proposals developed by outside organizations having the full range of facilities and skills at the system-of-systems level.

With the services' increasing emphasis in their visions of the future on high-level systems integration, traditional prime contractors that specialize in platform design and production have begun to offer architecture systems-integration services. Firms with core competencies in electronics and network-oriented activities are also angling for platform systems-integration work, arguing that interplatform integration (interoperability) is becoming ever more important in the design of the platforms themselves.

Prime contractors have focused for years on understanding the unique demands of the military customer, hiring retired military officers for important positions in their strategic planning departments. Private firms are also largely exempt from civil service rules, allowing them the flexibility to hire top technical talent when necessary;[141] for scientists who crave equity compensation, private firms can also offer stock options.[142] When technical teams develop internal rapport that generates extra value from synergies or experience, private firms have incentives to support such built-up human capital. Managing technical personnel is a core competency of technology-dependent private firms, including defense industry prime contractors.[143]

However, platform systems integration and system-of-systems integration are not the same task, and it is not even clear that developing skill at one helps very much in developing skill at the other. Platform integrators may improve their performance through any of a number of different activities: repeated design or prototype development experience; production experience; and maintenance of close relationships with applied technical laboratories, basic science research establishments, academic institutions, or the operational user community.[144] Their unique advantage is in linking systems-engineering capability with intricate knowledge of the manufacturing process, allowing them to take advantage of production efficiency advantages in the design process. Naturally, prime contractors emphasize the importance of production capability in their discussions of systems integration—just as military laboratories emphasize the importance of full-scale system testing. However, while this advantage surely carries weight, it is likely to be relatively small in the defense sector, where production runs are often short and very-close-tolerance production processes are often craftlike, minimizing the potential for major savings. Such production issues should consequently receive a relatively low weighting in the system-of-systems integration trade space, although system-of-systems specialists should strive to consider platform makers' concerns when they do their overall analyses and define their requirements. System-of-systems concerns about platforms' interfaces with the network should take precedence in transformation planning and acquisition.

Moreover, the potential for conflicts of interest—or at least for the *appearance* of conflicts of interest, a more stringent standard that has been deemed appropriate for government organizations—mandates a separation between architecture-systems integration and production in the defense industry. Production prime contractors have the technical capability to scan subcontractors' products, including the offerings of innovative commercial firms, for likely partners in the network-centric defense industry—that is, they can fulfill one of the key technical and management requirements of a systems integrator. They also can make technical decisions about interfaces, network standards, and other requirements definitions; by vertically integrating so as to combine platform and components-oriented design and production organizations, large prime contractors might provide technical systems-integration services with minimal transaction costs. But the idea of expanding the roles of established prime contractors faces a crucial nontechnical barrier—lack of trust. Manufacturers certainly test their products before delivery to the customer, but the customer also needs an independent ability to verify product performance—just as military laboratories emphasize. In addition, the customer might reasonably fear that a manufacturer's trade-off analysis might be biased in favor of the sort of alternatives that the manufacturer is expert at making— or even (unintentionally) skewed by the production contractors' technical understanding of particular systems and solutions as opposed to others.

This problem was first manifest in the defense industry in a 1959 congressional investigation of the relationship between TRW's satellite and missile production businesses and the TRW-owned Space Technology Laboratory (STL), which played a technical-direction role in Air Force development and production projects—including some for which TRW had submitted proposals. Neither protectors of the government trust nor members of the defense sector that competed with TRW on those space systems contracts would accept the situation, even though no specific malfeasance was uncovered or even alleged. STL was essentially split off from TRW to become Aerospace Corporation, an independent, nonprofit, nonproduction, systems-integration specialist, later designated an FFRDC.[145] That organizational innovation, which spread with the establishment of other FFRDCs and the similarly organized "university applied research centers" (UARCs), allowed the military's acquisition organizations to outsource the technical advisory role during the Cold War in a way that was protected from conflict-of-interest scandals.[146] Some FFRDCs, like MIT's Lincoln Laboratory, specialize in particular kinds of military-oriented research (advanced electronics, in that case), comparable in some ways to the in-house military laboratories but more closely tied to frontier academic research. While the core tasks of various FFRDCs overlap to some extent, Aerospace Corporation (space systems), MITRE (air defense), and APL (naval systems) are the ones that specialize in architecture-systems integration.[147]

The historical strength of FFRDCs has been their reputation for high-quality, objective advice. Through flexibility in salary negotiations and their quasi-academic status, FFRDCs have been able to attract high-quality personnel. Their promise not to compete for production contracts and to provide equal access to all contractors while safeguarding proprietary information has given them unique, independent technical capabilities.[148] However, they have frequently been criticized as inefficient and relatively expensive; while leaders of FFRDCs frequently claim that their nonprofit status allows them to charge less than a hypothetical technically equivalent, for-profit technical advisor, many others (notably leaders of for-profit firms, like SAIC) allege that the lack of a profit motive in FFRDC work leads to inefficient performance and the potential for featherbedding.[149] Legislation currently limits the budgetary resources available to FFRDCs and prevents the military from establishing any new ones.[150]

For-profit, nonproduction firms might be able to offer the benefits of FFRDCs while avoiding the controversies linked to nonprofit status. Small engineering companies like SYNTEK can offer technical advice to the military with a credible promise not to engage in production, but it is difficult to imagine such a firm nurturing a major laboratory with an independent research capability and agenda, at least under current procurement rules. Without direct access to such scientific assets, it is reasonable to question the ability of a consultancy to maintain top-level system-of-systems integration skills.[151] Larger for-profit firms like SAIC—which owns Bellcore, the former research arm of the Regional Bell Operating Companies (a partial descendant of Bell Laboratories)—offer to fill this niche, but to cover the overhead cost of such laboratories they resist pressure to abstain from all production work. Although for-profit firms in the defense industry have learned to form teams to develop major systems and sometimes even to join a team on one contract with a firm against which they are competing for another contract, real questions persist about how much proprietary data the for-profit contractors are willing to share with one another. A promise not to engage in production would allay some of the fears that prevent platform firms from becoming architecture-systems integrators, but major for-profit advisory firms are still limited by customers' and competitors' skepticism about their true, long-term independence.

NCW and Systems Integration: Performance Metrics

Specific metrics for comparing systems integration capabilities have not yet been defined in detail, so project managers may have difficulty selecting sources for technical advice and deciding how much investment in up-front systems-integration work is enough. Carnegie Mellon University's Software Engineering Institute (SEI), a research FFRDC, has developed a rating system for several information technology–related skills, including software engineering and systems engineering. The ratings assigned according to the SEI "capabilities maturity models" are based on a business's commitment to follow certain procedures designed to manage complex projects. Specifically, they emphasize maintaining control of documentation and interfaces to ensure systemwide performance as components and subsystems are improved in parallel. These software-oriented procedures are at least related to the broader systems-integration task, and they may provide a useful model for further work defining metrics for overall systems-integration capabilities.[152]

For the purposes of this report, however, such detailed metrics for evaluating systems integrators are not necessary. The key question in the systems-integration sector, as in shipbuilding and unmanned vehicles, is whether the transformation to network-centric warfare requires sustaining or disruptive innovation. If network-centric warfare builds on established performance metrics, established SI organizations will be able to implement transformation; if new performance metrics must be applied, then new systems organizations will be called for. Four systems-integration performance metrics require the attention of transformation advocates.

Technical Awareness. The bedrock of systems integration is familiarity with the technical state of the art in the wide range of disciplines that contribute to the components of the system. Systems integrators must be able to set reasonable, achievable goals for the developers and manufacturers of system components even as they "black box" the detailed design work for those components. If one component maker has a problem that it can solve only at great expense but that could be solved much more easily by changing the requirements of a different component or by altering the interface standard in a way that would cost other component manufacturers less, it is the responsibility of the systems integrator to understand and implement the necessary trade-off in the various component specifications. The more access the systems integrator has to technical knowledge of subsystems, the better it will be able to perform that role. There are many ways that a systems integrator can obtain this technical knowledge, including systematically and continuously training and educating critical engineers, hiring personnel from subsystem contractors, and seconding employees to other organizations to work in all phases of component design and production.

Transformation is unlikely to change the role of technical awareness as a systems-integration performance metric. To the extent that network-centric warfare draws on unfamiliar component systems, it may strain the technical awareness of established SI organizations. For example, emerging unmanned vehicle technologies may take over a number of tasks previously assigned to manned systems, requiring systems integrators to be familiar with the state of the art in UV technology if they are to make trade-offs between manned and unmanned systems. However, the systems integrator need not have the capability actually to design and build either the manned or the unmanned systems. The specific technical knowledge is not the core competency for the systems integrator; the sine qua non of systems integration is, instead, the ability to gain *access* to that knowledge, by working with subsystem contractors, academic experts, or in-house researchers.

Developing new sources and kinds of technical awareness may be the core competency of a systems integrator, but it is only natural that the less familiar the component technologies of a particular project are to a systems integrator, the less effective that integrator will be. Even the organizations with the broadest architecture systems-integration capability have specialties—Aerospace Corporation in space systems, for example, or MITRE in command and control. It is not obvious, however, that network-centric warfare demands new specialties. Instead, it seems to involve the advanced application of a combination of established ones—for example, reliance on space systems for surveillance and communications relay, on intensive exploitation of command and control networks and battle-management computation. If a new focus on the network characterizes the systems-integration task for network-centric warfare, MITRE, APL, and for-profit firms like Logicon and SAIC appear to have the necessary technical awareness. Perhaps the Software Engineering Institute's foray into integration provides the basis for a transition from a pure research FFRDC into a research and systems-integration combination (akin to APL) that specializes in network technology.[153] Although the commercial Internet has burgeoned well beyond its defense origin, the Arpanet, the original DARPA program has been cited as a classic example of the military's "systems approach" to advanced technology.[154]

The organizational framework through which established organizations' specialties should be applied to the new problems of network-centric warfare remains, however, an open question. Various systems integrators might offer competing technical proposals, each offering its best system solution to network-centric warfare challenges and pointing out flaws in alternative proposals. American pluralist government is built on the principle that the clash of ideas yields the best policy solutions; that clash of ideas might help compensate for each existing organization's implicit biases in favor of its technical specialties. APL might point out any pitfalls of Aerospace Corporation's space-based solutions, while Aerospace could illuminate the risks of APL's hypothetically bandwidth-consuming approach. Still, it remains the responsibility of the customer/buyer to evaluate competing claims in order to make decisions in the corporate interest of the Navy or, better yet, of the U.S. military as a whole.

Alternatively, a team combining the relevant technical groups from the established systems integrators might be able to offer a comprehensive technical base for network-centric systems integration. Ten FFRDCs and national laboratories combined to provide technical support to the Ballistic Missile Defense Organization through a teaming arrangement called the Phase One Evaluation Team (POET).[155] A full evaluation of the technical performance of the POET is beyond the scope of this report, but some preliminary observations are relevant. On the one hand, the POET clearly provided access to an exceptional breadth of technical talent.[156] On the other hand, the participant organizations retained their traditional customers, missions, and cultures; they may not have invested their best resources in, or devoted their full attention to, the missile defense effort.[157] A systems-integration team for network-centric warfare would gain similar advantages and face similar limitations.

To apply the full resources of the established systems integrators to the new challenges of network-centric warfare, it might be advantageous to create a new systems integrator with a new bureaucratic identity. But it would not be necessary to create such an organization from scratch; it would be very costly to replicate the investment in human capital that has already been made by established organizations. When MITRE was created as the systems integrator for the SAGE air defense system in the late 1950s, its core was formed from Division Six of Lincoln Laboratory, which chose at that point to focus on research rather than systems integration. MITRE then proceeded to expand its technical awareness into new areas, integrating air defense missiles like the BOMARC into an air defense system initially designed to cue fighter interceptors.[158] Today, it might be possible to blend various technical groups spun off by the established organizations and form a new FFRDC. The new institution would maintain the well-understood core competency in nurturing technical awareness but would do so in the service of a new customer and organizational mission.[159]

Each of these three candidate organizational forms to supply systems integration for transformation—competition among architecture integrators, a team of architecture integrators, or a new architecture integrator—relies on the built-up skills of established institutions; they are evolutionary changes required to proceed with sustaining innovation along the technical awareness performance metric. The financial ownership structure of the technical advisor is less important than its underlying skill base, which can be derived from existing systems-integration groups.

Project Management Skill. Efficiency has rarely if ever been the only goal of military acquisition programs. In addition to serving economic goals, the projects need to meet military requirements and to satisfy political constraints.[160] Nevertheless, efforts to control costs have been a continuous feature of defense policy. Warfighters would always like to acquire more systems; technologists always can use additional resources to push the performance envelope further; and politicians always have nondefense priorities, including lowering taxes. Because all three groups also try to plan their expenditures as part of the budgeting process, they need project cost and schedule estimates that are as accurate as possible.

For complex acquisitions with numerous, heterogeneous components—a system of systems—reliable estimates are difficult to come by, due to the vast amounts of information that must be managed to describe the current and projected state of progress. Participants also have incentives to hide certain information from oversight. Sometimes they believe setbacks to be temporary (that they will get back on schedule, the promised performance trajectory, or the estimated cost projection before they have to report problems); sometimes they fear that full disclosure will aid competitors or lead to pressure to renegotiate fees and expropriate profits. Managers learn to report data in favorable ways, rarely involving real malfeasance, that can give a biased picture of progress that protects ongoing projects from scrutiny.[161] They also enthusiastically embrace acquisition reform efforts and management fads that promise to reduce costs in the future—after enough investment has been sunk into the project to lock it into the political landscape, whether or not the efficiency benefits of the reform ever actually materialize.[162]

System-of-systems integrators have the expertise to manage projects as well as possible in the face of these constraints. The better a given systems integrator performs in that project management task—setting accurate schedules, projecting attainable technical goals, and minimizing transaction costs among the many organizations that have to contribute to a systems contract—the greater the incentive the buyer has to hire that systems integrator. Project management skill is a key performance metric for SI organizations.

Transformation calls for sustaining innovation in project management. Ultimately, for network-centric warfare to be useful to the warfighter, a number of different programs (for example, ships, aircraft, unmanned vehicles, munitions, and sensors) need to deliver compatible systems to the fleet in the correct order; the schedules need to be timed so that the various deployment dates form the network. Cold War programs like the Polaris fleet ballistic missile program, which required tremendous innovation in missiles and guidance, in communications and navigation, and in submarine platforms, faced the same sort of management and scheduling problems. System-of-systems integration was effectively invented precisely for the purpose of managing such massive, heterogeneous acquisitions.[163] Network-centric warfare may require integration of an even broader array of components, making the system-of-systems integration task even more difficult. But systems integrators are already applying modern information technology to manage complex subcontractor networks, to scan for technological leads that might contribute innovative solutions to military problems, and to interact with potential new suppliers, innovating to support this core task.

At the platform integration level, the project management task under transformation will be little changed from its previous incarnations. Whether any given platform integrator is well positioned to participate in transformation will depend on the demand for its technical skills—whether network-centric warfare calls for sustaining or disruptive innovation in that sector of the defense industry. The platform integration task will continue to include management of subcontractor relationships and the detailed design of military systems. In sectors dominated by sustaining innovations, platform integrators' databases of successful subcontractors and procedures for working with the social and political constraints of the government contracting environment will contribute to successful acquisition programs. Despite acquisition reform advocates' appropriation of phrasing from transformation advocates—the "revolution in acquisition affairs" or "revolution in business affairs"—the quest for acquisition reform is separate from military transformation.

At the architecture systems-integration level, transformation's biggest challenge in project management will stem from the need to integrate the plans and schedules of several powerful customer organizations. The mechanism by which a technical direction agent for network-centric warfare can assert control of the technical aspects of project management may change (changes in the customer relationship will be discussed below in the section on customer understanding). But the core project management task will not change much; system-of-systems integrators will have to integrate some new technical tasks into military systems development, but the disruptive innovations, if any, will fall at the platform or component level rather than that of organizational and management techniques for the system-of-systems project. Transformation requires that high-level systems integration evolve along a familiar performance trajectory, contributing as much efficiency and scheduling accuracy to major systems acquisition as possible. The sustaining nature of that innovation suggests that transformation will not change the core composition of the system-of-systems integration sector.

Perceived Independence. The key role of a system-of-systems integrator in defining the technical requirements of various system components (and hence of the system as a whole) requires that it be able to make trade-offs in the interest of system performance rather than in the interest of the organizations that design or make the system. The architecture-systems integration task is tremendously complicated; military systems have multiple goals—peak warfighting performance, sustained political support for the acquisition program and for the national security strategy, and minimal expenditure of resources for acquisition, maintenance, training, and operations.[164] That complexity, along with the requisite technical expertise, effectively guarantees that detailed decisions in system-of-systems integration will not be completely transparent to military customers, congressional appropriators, or the defense industry primes and subcontractors that supply components of the system. All of those groups must trust that the systems integrator has considered and protected their interests in making its architecture-definition decisions; any organizations that feel that their trust has been violated are in a position to create a scandal by complaining publicly. They are constrained, however, by the understanding that complaining too often or too loudly can subvert the entire process of providing for the national defense. They cooperated in the Cold War evolution of system-of-systems integrators in ways that minimized the problem of bias in system definition, and that lack of bias is today a key performance metric for system-of-systems integrators.

The difficulty in maintaining independence for architecture systems integration is compounded by the pecuniary incentives in defense acquisition. Like all organizations, systems integrators have an incentive to favor solutions that maximize their own organizational rewards, maintaining and exploiting their positions as key nodes, connecting customers and producers in the organizational network of the military-industrial complex.[165] This bias may be purely tacit, such as that by which scientists propose certain types of technical solutions based on their particular expertise, thereby reinforcing the value of that particular expertise. It may also be structural: profits in the defense industry have disproportionately accrued to production rather than research or technical advisory organizations, in large part because profits are regulated, formally and informally, at a certain percentage of project revenue, and the bulk of the acquisition spending is concentrated during the procurement rather than the systems development phases.[166] Further, in the post–Cold War threat environment, wherein the United States faces no peer competitor, firms having "critical masses" of workers (generally production rather than technical organizations) have been able to add considerable political weight to their pleas for financial support from congressional appropriators.[167] Consequently, the financial prospects for pure system-of-systems integrators are weak, and they face pressure to incorporate systems integration vertically into production capability. Freedom to choose optimal technical solutions is constantly threatened at the margin by the bureaucratic interests of the services and the political power of platform producers. Because this pressure is well known, trust from the customers that the systems integrators will protect the military's interests and not simply their own material interests, is also threatened.

Bias is built into the very makeup of most established systems-integration houses. They originally served particular customers, and the needs of that customer were well known. Lack of bias in this context meant that within their own issue domains they might reasonably be expected to play the honest broker. In turf battles with external forces, however, they might favor particular types of solutions. Thus Aerospace Corporation might be unbiased in telling the Air Force about how to organize and equip its own space capabilities, but less so when arguing for space-based solutions as opposed to nonspace-based approaches proposed by other government entities.

By and large, the FFRDC/UARC system of nonproduction technical advisors functioned successfully during the Cold War.[168] The contractual relationships of FFRDCs and UARCs with the government commit them not to engage in production. Some degree of tension inevitably remains between the producer firms and the FFRDCs, who insist that they need to engage in prototype building that is quite similar to production in order to maintain their SI skills. Such tension seems particularly likely to escalate in the software industry, where the development and production phases of a code-writing project frequently overlap.

APL, for example, has gotten into trouble for mixing production with systems integration, specifically in the current dispute over the best technology for the Navy's Cooperative Engagement Capability. Solipsys, a software firm founded recently by disenchanted former employees of APL, has created a rival system, the Tactical Component Network (TCN). Solipsys claims that the Navy has not given it a fair hearing, at least in part because APL is both the technical advisor to the Navy and the developer of CEC. Regardless of the technical merits of CEC versus TCN, and here opinions vary widely,[169] the controversy would be less bitter if APL were not exposed to charges that it favors one solution over the other because it developed that alternative and would participate in its production. The Navy, which will have to decide between the two approaches for its Block 2 acquisition of CEC in 2004, has a real problem evaluating the technical claims of the competing organizations.[170] Even if the Navy finds a way to make the technically correct decision, conflict-of-interest claims will arise—as they already have—and the likely outcome will be extra oversight of the CEC program, increasing costs and undermining political support for that key early procurement step in developing the Navy's common operational picture, which is required for network-centric warfare.

Scandals alleging "waste, fraud, and abuse" and cost and schedule failures have derailed military investment in the past, and conflicts of interest might be a threat to the Navy's move toward network-centric warfare. The peaks in the major cycles of the U.S. Cold War defense budget were associated with procurement scandals, which at least superficially played a role in reversing the defense budget trend. Even if, at the time, structural factors like the changing threat environment or the completion of a generational change in key equipment were bringing the procurement cycle to an end, calls to rein in defense-acquisition abuses were a proximate cause of the downturn in the defense budget.[171] The Future Years Defense Budget now calls for a major increase in procurement spending for the next several years—the defense budget's new cycle. If the military leadership hopes to use that spending to develop and acquire the systems to implement transformation, the cycle must not end prematurely in scandal.

Military transformation relies on sustaining innovation to meet the metric of "perceived independence" for system-of-systems integration performance. Some of the Navy's technical advisors for whom lack of bias was a key core competency during the Cold War have actually begun to stray from that trajectory, under pressure to defend sunk investment in particular technical approaches or to increase revenues by exploiting "industrial funding" privileged by recent acquisition reform. Those post–Cold War pressures might have been disruptive; the 1990s cutbacks among the nonproduction technical advisors may reflect the effects of disruptive innovation that over time would have revamped the industrial base for system-of-systems integration. The best way to implement network-centric warfare would be to return to the well-known "lack of bias" performance trajectory as soon as possible, while suitable organizations still exist with core competencies to proceed with system-of-systems integration.

Customer Understanding. The Navy, with all its communities (primarily the three led, respectively, by aviators, submariners, and surface warfare officers), is a complicated organization with a long institutional history, unique traditions, and organizational biases developed over generations of operational experience. More formally, there is a large body of strategy, tactics, doctrine, and training processes that distinguish the Navy from the other services, from other government agencies, and from the private sector. The other services and supporting intelligence organizations have similarly developed organizational identities and perspectives on warfighting and national security strategy.[172] The success of each system-of-systems integrator depends on how deeply it understands the naval and military environments, because the integration organization's architecture definitions and project management decisions must serve its customer's true goals, which can be difficult to articulate in a simple, program-specific, "statement of objectives." Navy-oriented systems integrators (for example, APL, SYNTEK, and Lockheed Martin Naval Electronics and Surveillance Systems) have built up a great deal of tacit knowledge about how and why the Navy operates, knowledge without which they would not be trusted to perform system-of-systems integration. Customer understanding is important for any organization, but it is a uniquely vital performance metric for architecture systems-integration organizations.

Customer understanding is a moving target; long experience alone is insufficient. A systems integrator must commit itself to invest continuously in its military-operational knowledge base. It must monitor lessons learned from recent exercises and operational deployments, as well as changes in military doctrine and national grand strategy, in order to maintain the "right" kind of technical awareness. Ideally, members of the SI organization should participate in war games and exercises wherein the Navy tests new operational concepts and introduces virtual prototypes of future platforms and subsystems. Teaming in various forms can only help personnel and organizations develop a greater appreciation for mutual idiosyncrasies. A large part of customer understanding is the maintenance over time of interorganizational relationships that transcend individuals and projects.

Unfortunately, customer understanding might also reinforce institutional inertia and reify the status quo; in many ways, it is analogous to bureaucratic "capture," as a result of which a regulator sees things from the perspective of industry rather than the public interest. Yet these dangers are not best avoided by creating firewalls or by artificially introducing change from the outside. Rather both the customer and the SI organization must self-consciously distinguish between customer understanding for the sake of overall success and collaboration for the sake of blocking change or protecting institutional interests. In short, the systems integrator must be free (and protected) to resolve trade-offs in ways that may harm short-term customer interests but guard the long-term health of the organization as a whole.

The need to make trade-offs and provide analyses of alternatives that may threaten existing programs and the short-term plans of system-of-systems integrators' customers puts SI organizations in a delicate position. Individual services are wary of criticism and fear losing ground in budgetary competition with other services, just as individual platform makers may resent the oversight that an independent systems integrator exercises on particular projects, even while understanding its necessity for the overall success of national defense investment. System-of-systems integrators' customers must be confident that the systems integrator has their true interests at heart.[173]

At the architecture systems-integration level, transformation's biggest challenge lies in the fact that the system-of-systems crosses many organizational boundaries. This requirement is especially severe in the more expansive visions of transformation that emphasize network-centric warfare as a joint rather than a service undertaking. Communities within the services have strong, independent identities, ideas about how wars should be fought, and priorities for setting schedules and allocating funding. Each service, in turn, tries to influence the course of transformation—and to influence the definition of the system-of-systems by pushing preferred definitions of the trade space and by defending and funding particular programs that the overall systems integrator must then integrate into the network-centric force structure. Architecture systems integrators will have to understand and balance the conflicting motivations of the several customers, most of which have great difficulty incorporating multiple goals into their organizational identities.[174] These considerations suggest that a truly joint systems approach may require establishment of a single, joint acquisition agency to which a single system-of-systems integrator could be attached. However, added organizational layers between integrators and their service customers, who will actually operate military systems, might degrade customer understanding, reducing the effectiveness of analysis of alternatives. Adopting a single buyer for transformational systems might also threaten the diversity of approaches that interservice rivalry could otherwise provide.

Service visions of transformation will require that system-of-systems integration organizations pursue sustaining innovation in customer understanding, building on established communications channels to the fleet, doctrine developers, and the acquisition community. System-of-systems organizations now at home serving particular subsets of the military may have difficulty developing contact networks and perfecting customer understanding at the "higher" level system-of-systems integration environment—for example, a firm supporting only space systems that finds that both space and terrestrial systems now need to be analyzed as alternatives within the network. A meaningful, joint transformation vision will require a more disruptive innovation trajectory for system-of-systems integration in the way that jointness routinely requires disruptive innovation, squeezing out established organizations and suppliers. However, much as established architecture systems integrators have the skills to expand technical awareness into new areas, those organizations also have the skills to focus on developing customer awareness as a means of staying in business. Transformation does not change the organizational goal of customer understanding, but organizational boundaries will be at least as difficult—likely more difficult—to overcome than interdisciplinary boundaries in technical awareness.

Sector Evaluation

Transformation relies explicitly on intense interoperability, one of the key components of system-of-systems integration; accordingly, transformation and systems integration have become tied together in a very public way. At this early stage of transformation, however, another component of system-of-systems integration is even more important—trade-off studies to establish the objectives and requirements for the component systems that will be acquired as nodes and network elements.

Certain established systems-integration houses, like APL and MITRE, clearly have expertise that is closely related to present plans for network-centric warfare, and they should play major parts in the network-centric defense industry. Similarly, some of the production-oriented prime contractors have high-level systems-integration groups that on technical awareness and project management grounds might join the nucleus of competitive SI suppliers. However, in the face of commitments to sustaining innovation in terms of the lack-of-bias and customer-understanding performance metrics, prime contractors' skills are more likely to be optimally applied in the service of platform rather than in architecture systems integration.[175] Given the predominance of sustaining innovations in the systems-integration sector's part of transformation, the key step in preparing the defense industrial base for network-centric warfare is not to try to change the cast of characters but to update and focus the technical emphasis of the Navy's own acquisition community.

There is no reason to invite platform-making prime contractors into the systems-integration sector as part of transformation. The primes want in because they perceive that systems integration is "where the money is," at least in the short term; they perceive it as the locus of greatest responsibility in the future defense industry. Moreover, as political pressure builds in support of transformation, and systems that are not perceived as transformational (like the Army's Crusader self-propelled howitzer) become vulnerable to cancellation, prime contractors are looking for ways to link their activities to transformation. The logic for the primes is the same as it always has been: if a particular kind of acquisition reform is popular, your programs should be "demonstrators" of the new technique; if systems analysis and PERT charts are the way to show budget and schedule control, your programs should use them; if network-centric warfare is the future operational concept, your programs should emphasize their connectivity.

Acting as a systems-integration agent might be the best protection of all for a prime contractor's business base. Production firms in the defense sector might be expected to complain about outside systems-integration houses' role on particular projects, because the advisor's job includes raising awkward criticisms of the prime contractors' technical approaches and production skills. One way to avoid such criticism would be to make systems integration part of the prime's job. However, given the importance of independence for quality systems integration and the fact that up-front technical advice and coordination will help to keep transformation programs on schedule and budget, production contractors should find it in their interest to support outside SI organizations (especially if paid mostly from the military infrastructure budget rather than from specific projects' budgets).

On the other hand, it remains very difficult for the Navy to choose its technical advisors for system-of-systems integration, because systems-integration performance metrics are difficult to operationalize and tie to the traditional framework for defense contracting. No top-down metric that is developed for systems-integration skill will be able to substitute for organizational competition. The various systems-integration organizations can offer a diversity of technical approaches and system-of-systems proposals, and they can offer technical commentary on and critiques of each other's proposals, giving the military customer enough advice to make informed choices early in the transformation process. The consolidation of the defense industrial sector through mergers and the reduced post–Cold War demand for long production runs has limited competition for production contracts; the overhead cost of maintaining multiple production lines for each weapon system is also unacceptably high, even in the current defense budget environment. However, competition among technical advisory organizations—each with a different design philosophy or technical focus—is relatively inexpensive to sustain, and those dedicated systems integrators should be able to help monitor technical efficiency during the production phase of the acquisition process. Meanwhile, in competing for their shares of the technical advisory role during the upcoming military transformation, these organizations will monitor each other's performance, point out technical flaws in competitors' proposals, and help decide how and how much to invest in systems integration. Exploiting competition among dedicated SI organizations should be a relatively low-cost response to the tension between budgetary pressure and the high resource demand of investing in military transformation. In the end, however, the buck must stop somewhere. Competition among systems-integration organizations may keep everyone honest and allow ideas to be triaged, but the Navy itself must sort through competing claims and make decisions.

Major acquisition projects or groups of related projects during the Cold War often spawned new procurement and advisory organizations. A new acquisition organization/systems integrator partnership might now facilitate the Navy's transformation effort. Advocates of network-centric warfare frequently note that the current acquisition system is organized on a platform-by-platform basis, which naturally deemphasizes crucial network investment. The potential problem is very much akin to the barriers to investment in missile defense through traditional acquisition channels that led in the 1980s to the creation of the Strategic Defense Initiative Office, predecessor of the Ballistic Missile Defense Office. The Navy should consider giving network-centric warfare a similar home in a new acquisition organization that will develop a bureaucratic interest in acting as the budgetary advocate for transformation. Because the network is intended at least to link systems from all of the communities within the Navy, this new organization would report directly to the highest echelon of Navy acquisition decision making, the Secretary of the Navy.

The new organization could also take responsibility for supporting a new technical advisory organization that will develop expertise specifically in the network and node requirements for the Navy after Next. This organization will, in all likelihood, borrow personnel and even intellectual capital (for example, lessons-learned databases) from existing systems integrators as well as develop new competencies necessary to handle the complexities of the network-centric environment. Any such new systems integrator would need a high-level sponsor, a reasonable budget, insulation from the inevitable bureaucratic infighting, and, most of all, time to develop the trusted relationships and track record of success that characterize systems-integration houses. The political pressure behind transformation may not be able to wait that long. In the case of the Reagan-era surge in funding for missile defenses, a new acquisition organization was created because the bureaucratic identities of the services' systems commands diverted their efforts from missile defenses into traditional systems; however, technical support for the missile defense systems' diverse components fundamentally relied on systems-integration skills that were available from established organizations. As a result, the POET, composed of the established systems-integration houses, provided effective technical support.

In the current policy environment, the balance is tipping away from dedicated systems-integration houses like FFRDCs and the technically skilled professional service corporations, and toward prime contractors that build platforms. If the military services succeed in reversing that trend and creating a POET-like team for network-centric systems integration, perhaps that should be considered enough of a victory. It would provide at least minimal protection from scandal that might derail the information-technology revolution in military affairs. Despite the questions that some have raised about whether the POET optimized technical support for missile defense, a POET-like team for network-centric warfare might well make important strides toward improving the technical future of the American way of war.

Conclusion

CHAPTER FIVE

Our conclusions and recommendations, based on nearly two years of research and hundreds of interviews with defense industry executives, government officials, policy analysts, and scholars, focus on (1) the defense industrial implications of military transformation and (2) how to ensure industry support for transformation. In both areas we make recommendations to help the Navy and DoD as a whole achieve transformation in partnership with both Congress and industry. Our intention here, as in the rest of the report, is not to advocate transformation per se. Rather, we seek to examine the conditions under which transformation might be successful when, and if, the U.S. Navy, the other military services, and the Department of Defense seriously commit themselves to adopting and implementing a new vision of warfighting.

The Defense Industry and Military Transformation

Many of our findings run contrary to recent analyses of military transformation and of the current and likely future state of the defense industrial base. We do not find persuasive most of the mantras of transformation advocates. The defense industry is not going to disappear. Commercial information technology firms will not displace defense sector primes as the major suppliers of equipment and expertise to the next military or the military after next. Innovation will proceed apace with or without significant commercial-sector participation, providing the military can decide upon the goals of transformational innovation. Systems-integration organizations will be able to translate goals into requirements, and platform integrators will be able to develop and produce the detailed equipment designs.

Nor do we share many of the current concerns about the health of defense firms. Recent anxiety about the profitability, or lack thereof, of defense firms were more an artifact of the "dot-com" bubble and the first crest of the "new economy" than evidence of problems within the defense industry itself. While some firms have exited the defense business, others have focused more closely on defense; such firms have done remarkably well for their shareholders in the last year or so. While access to international markets could be improved, the simple fact remains that the U.S. R&D and procurement budgets are by far the largest prize in the global defense business. If anything, foreign defense firms will have trouble surviving without access to the American marketplace; American defense firms are more able to survive, if not thrive, in a fragmented international marketplace. Neither profitability nor globalization is a key issue or challenge for the implementation of transformation visions.

At the highest level of generality, we do not believe that military transformation will require wholesale defense industrial transformation. Traditional defense suppliers have provided the United States with military capabilities unparalleled in world history; they will continue to do so far into the future. Calls to purge the term "*defense* industrial base" from our lexicon in favor of simply "industrial base" do not take account of the unique characteristics of the firms that provide weapons and systems to the U.S. military. Defense firms have numerous competencies—from experience interacting with the military culture to the administrative infrastructure necessary to meet unique government regulations—that are not found elsewhere within the American economy. Simply having the administrative mechanisms in place to deal with the complexities of Defense Federal Acquisition Regulations, for example, constitutes a core competency that is not, and cannot be, duplicated by most potential commercial entrants into the defense business. Even acquisition reforms will not change the importance of these competencies; the so-called revolution in business affairs will only go so far toward making federal contracting similar to its civilian counterpart. Congressional representatives and, indeed, the federal workforce as a whole demand a higher level of transparency, fairness, and accountability than is fundamentally compatible with standard commercial business practices.

This is not to argue that no changes within the defense industrial sector or the government's acquisition system will occur. In some specific niches, nontraditional suppliers will play roles in military acquisition in the future. They are highly unlikely, however, to displace General Dynamics, Northrop Grumman, Lockheed Martin, Boeing, or the other established prime contractors. On the government side, there also may be a great deal of tinkering at the margins—with changes in the Advanced Concept Technology Demonstrations system, export controls, and dual-use technology regulations, for example—perhaps even a major acquisition reform effort by Congress. The DoD leadership has pushed for numerous reforms and aggressively advocated a revolution in business affairs to reform both its own internal procedures and its relationship with the private sector. In the end, however, the basic premises that have led the United States to today's acquisition system, however Byzantine, remain valid.

A principal finding of our research is that it is less helpful to discuss the defense industrial base as a whole than it is to focus on specific sectors providing particular types of capabilities and the proposed roles for particular sectors in transformation visions. Many of our findings are sector specific.

Shipbuilding

The network-centric vision, if fully realized, suggests that it is not simply networks but the nature of the nodes (read "platforms") that must change. If the NCW vision is adopted in its most robust form, the Navy should soon be buying smaller, less complex ships that are designed to operate in a highly complex, fully networked, system of systems. Warships will no longer serve as multipurpose vessels equipped to operate on their own. They will instead be most effective as specialized components of much larger systems.

This vision must be tempered by two realities. First, large, multipurpose warships are unlikely to disappear from the fleet any time soon. Legacy systems, from nuclear-powered aircraft carriers to DDG 51 destroyers to SSNs, will remain in the fleet for several generations. Further, most transformation advocates acknowledge that their views do not envision changing the nature and composition of the entire military. Often drawing on analogies to the transformation achieved by Nazi Germany's armed forces that enabled the adoption of blitzkrieg tactics, they suggest that roughly 10 percent of the current force needs to be transformed. Second, potentially innovative shipbuilding programs, even if they focus on affordability, remain very expensive, a factor that will limit the rate of deployment of new ship classes.

Christensen's innovation framework suggests that some nontraditional suppliers will enter the defense industrial base—depending, of course, on the specific performance metrics the Navy adopts as it winds its way through the acquisition cycle for programs like the DD(X), especially the LCS component of that program. Yet this possibility may be undercut by the nature of the relationship between the Navy customer and the potential shipbuilding competitors. As in all defense industrial sectors, established suppliers enjoy many advantages over their commercial and international competitors. They have long-standing and relatively successful business relationships with the Navy and the other services dating back for decades. They have invested heavily in understanding the needs of the Navy, both by hiring retired naval officers and by closely monitoring the decision-making processes in Washington. Finally, they maintain large and active lobbying organizations to ensure that the obvious benefits of preserving existing firms (and their facilities) remain at the forefront of public debates. All this suggests that even if it makes sense for firms like Bender, Bollinger, or Halter Marine to participate in the production of transformational naval ships, they would in all likelihood do so in partnership with larger shipyards or other platform integrators.

UAVs

Network-centric warfare relies heavily on UAVs both as nodes (e.g., combatants, in the form of UCAVs, and carriers of various sensor suites) and as parts of the overarching net (e.g., communications relays) linking various commands and components together into the "global information grid." Even more than in shipbuilding, it is impossible to identify the performance metrics that will establish what constitutes a successful UAV and which firms can be expected to produce such craft. Most experts with whom we discussed UAVs were unable to articulate what distinguishes UAVs from other possible ways of accomplishing various missions or what makes one UAV design better than its competitors. Many of our informants, clearly, had not thought through performance-metrics issues.

Based on our efforts to understand the emerging UAV sector, there again seems to be little reason to believe that new firms will suddenly develop disruptive technologies—say, propulsion systems or aircraft control systems—that will allow them to attract new customers or undermine existing products (and their producers). There is a fairly large number of firms developing and producing UAVs today (and even more that have recently produced prototypes and demonstrators) that are already firmly ensconced in the traditional defense industrial sector; the list includes Boeing, General Atomics, and Northrop Grumman's Ryan Aeronautical. To a greater or lesser extent, each of these firms (with its second-tier competitors, such as AAI) is currently producing sustaining innovations and scanning the technology horizon for firms or ideas that can improve the performance of its products. It seems unlikely that they will be surprised by the work of start-ups. Even if they are, a rational response to technological surprise would be to acquire or license the new technology and incorporate it into proven manufacturing and marketing systems.

Yet the possibility of disruptive innovation remains, because the performance metrics for UAVs are neither well established nor well understood. In theory, over time, new standards that undermine today's market leaders might emerge. This observation may be especially true with regard to the Navy; one potential upside to the Navy's off-and-on relationship with UAVs is that the door remains open to new firms and new technologies. The Navy has not yet settled on one particular course. Further, because naval UAVs (at least those designed to be deployed with the fleet) may have unique performance constraints—launch and recovery systems or marinization, for instance—this portion of the sector may be ripe for disruption.

Moreover, some of today's UAV manufacturers have had rocky relations with their naval customers. General Atomics has been unsuccessful in its efforts to market its Predator series to the Navy; the firm's officials suggest that this failure stems from Navy hostility to UAVs in general. Northrop Grumman's Global Hawk has been plagued by technical glitches and cost overruns. In short, the relatively long-standing customer-supplier relations that are in place may not remain entrenched if credible alternative suppliers emerge.

Systems Integration

If network-centric warfare or, indeed, any approach to future warfighting is to succeed, systems integration—rather than platforms and even networks, per se—must be a high priority, both organizationally and financially. There are many obstacles to investing in systems integration, however. In Congress, systems integration has weak political support, because systems integration projects do not employ as many people as do platform programs, let alone in such concentrations as in Groton, Newport News, or Pascagoula. Systems integration often also has weak support from industry, because, under traditional business models, industry profits come from production rather than front-end research and development or maintenance. Systems integrators may even have an adversarial relationship with platform builders on any given program (when they do their job right, systems integrators critique contractor performance and recommend trade-offs that threaten primes and their suppliers). When a systems integrator is also a prime or part of a vertically integrated firm with multiple divisions, the government should fear that it may not receive optimal solutions. As one of its competitors put it, "When you ask General Dynamics a question, you may well receive a General Dynamics answer."[176]

Buying adequate systems-integration expertise is a responsibility of the services and DoD during the transformation process. However, since it is difficult to find well-recognized metrics to choose high-quality SI organizations, it is difficult to decide how much funding for systems integrators is enough. As discussed above, a variety of firms and types of firms will play roles in systems integration; it may very well be that to get the level of systems integration they require, DoD or the Navy will need to create a new organization with systems-of-systems integration responsibilities, much as the Strategic Defense Initiative Office was created in the 1980s to overcome organizational obstacles to investment in missile defense. This system-of-systems organization should probably report to the Secretary of Defense or the Secretary of the Navy if it is to have the authority necessary to make integration a priority within the service. It must also have oversight over the wide range of Navy acquisition programs—from ships to UAVs to aircraft—in order to make and enforce key decisions regarding architecture, trade-offs, and interface standards.

If DoD or even the Navy itself creates a new acquisition organization for system-of-systems integration, in all likelihood technical and professional assistance from a private-sector contractor will be required. It would be difficult to stand up a new governmental organization with all the technical expertise implied by the performance metrics discussed in the systems-integration section of this report. The type of support organization that would make the most sense has yet to be determined, but our initial judgment is that an FFRDC or an organization of that kind would be a leading candidate. FFRDCs have long histories of success in supporting large-scale projects with heterogeneous components; more to the point, FFRDCs avoid, by statute and design, many perceived conflicts of interest, have the resources to hire skilled personnel, and can develop the requisite customer relationships and technical competencies.

Transforming the Navy

When we began our research, few industry officials—and perhaps even fewer naval officers—understood the meaning of network-centric warfare and its implications for the future of the Navy. Today, NCW and its associated concepts are part of the daily discourse. As our earlier review of the military vision debate demonstrated, the Navy, the military as a whole, and even Congress have increasingly accepted the concepts of network-centric operations. Yet getting the terminology right is not enough.

For naval transformation to succeed, the Navy itself must rally behind its transformation vision. Specifically, the various naval communities and commands must support NCW by making the resource allocation choices necessary to support transformation. If they do, firms will determine their future business strategies on the basis, at least in part, of these new clues about their customer's future acquisition strategy. Firms will not, however, focus their internal R&D investments, technology search patterns, merger and acquisition plans, or personnel decisions on transformation before their customer has committed itself to it. Military transformation must be customer rather than supplier driven.

Elements of the Navy that do not understand or believe in the potential benefits of NCW may pay lip service to the terminology but fail to shift their ideas, personnel, and funding. Their reluctance would impede transformation by encouraging stovepiped programs, allowing projects that do not embody NCW principles and performance metrics to continue, and generally undermining efforts to enact major changes in how the Navy does business. Faced with mixed clues, firms will have few incentives to reorient themselves. They will have more reason to continue existing programs and maintain their cozy relations with traditional customers.

Ideally, the needs of the current Navy, the Next Navy, and the Navy after Next must be deconflicted. Especially in the current environment of an ongoing global war on terror, immediate operational requirements often bump up against projects with payoffs that will be realized much farther out in the future. Such conflicts are so pervasive that they could retard the ability of the Navy to respond to calls for transformation. The readiness of today's fleet and programs for the next one thus threaten the foundation of the one after that. Tension among these three navies will undoubtedly be evident well into the future.

A large portion of the fleet in the Navy after Next will consist of the "legacy" platforms of today's Navy and the Next Navy; it must be able to work with the new platforms developed for the Navy after Next. Contractors like Raytheon are researching ways to improve connectivity among the various generations of platforms, weapons, and sensors that exist now and will soon join the fleet. Yet program managers often have trouble finding institutional sponsors to fund such projects. Such small-scale interoperability programs are often orphans that must be pushed by industry rather than pulled by the acquisition side. It is even more troubling, however, that no one—neither contractors, the Navy laboratories, nor the Navy systems commands—appears to be systematically thinking through the large-scale system-of-systems architecture questions facing an NCW Navy. For example, what, if any, systems-design problems are inherent in overlaying an expeditionary sensor grid over existing and proposed spaced-based sensor systems? What are the trade-offs inherent in "shifting complexity"—in NCW language—from nodes to the network?

At an even higher level of generality, there are also conflicts between technological optimists and observers more skeptical of the scientific and engineering promise of many projects proposed for the Navy after Next. The danger exists that senior decision makers will seek to reap the technological promise of the Navy after Next prematurely—making it impossible to maintain interface control and systemwide documentation, and diverting resources from the planning of an optimal NCW system. Others expect dramatic decreases in program costs based on a promised revolution in business affairs and dramatic gains in manufacturing productivity. Overselling may make established but not-yet-in-production programs for the Next Navy less able to survive test failures or developmental delays that cost the taxpayers money or the services time. Already the rush to deploy Predators and Global Hawks in Afghanistan before they had completed OT&E has proved to be a double-edge sword; while both platforms have demonstrated their usefulness, they have suffered failures that have led some to question the future of UAVs generally.

Program difficulties that lead to cancellations or sharply decreased funding (resulting in smaller buys or extensions) may increase the resistance of the traditional defense industry to transformation strategies. After all, General Dynamics' Electric Boat, among others, was burned by the abrupt cancellation of the *Seawolf*; virtually the entire range of Navy contractors felt betrayed by the Navy's management approach to, and ultimate cancellation of, the Arsenal Ship program. Today some grumble about the recent fate of DD-21 (though the official line is that the DD[X] family of ships will incorporate much of the work done for DD-21). As for transformational programs, several industry executives we met with argued that the Navy had botched its UAV programs by failing to move from prototypes to production, by demanding unreasonable performance from immature technologies, and by commissioning "studies" rather than buying real prototypes.

In view of past problems, naval transformation will require careful management of the Navy's political relationships with Congress and industry. Congress may well be reluctant to commit to new, potentially expensive programs in view of fiscal realities and the vocal service, industry, and public constituencies that surround existing programs. Even the increases related to September 11 have not relieved pressure to divert resources from investment for the future into current consumption by operational forces. In this uncertain budgetary environment, political, organizational, and bureaucratic strategies matter.

The Navy should be wary of overpromising the cost-saving benefits of acquisition reform, mergers and acquisitions, new manufacturing technologies, and various performance metrics for its systems. For example, when General Dynamics claimed that it could save some two billion dollars if it acquired Newport News Shipyard, many were rightly skeptical.[177] If projected savings from industry consolidations and management reforms do not materialize, the Navy will be forced to make unpleasant choices and may lose credibility needed in the future. Political deals could leave programs underfunded in the out-years and increase the possibility that a program will be canceled.[178] The same holds true for the reputed savings from UAVs and platforms with reduced manning. Already the military has seen the cost of UAVs escalate,[179] and personnel removed from a DD(X) destroyer or the LCS may simply be assigned elsewhere.

Linking transformation to operational requirements—from changes in the strategic environment to changes in potential adversaries' technological sophistication or military preparations—will also help maintain momentum toward transformation.[180] Indeed, this appears to be part of the Bush administration's overall strategy. Programs and reforms that will presumably aid in the war on terror are deemed "transformational" and are thus popular. This dynamic is especially clear with regard to UAVs; supporters have hailed UAV successes and downplayed their failures both to increase funding and to overcome cultural resistance within some segments of the military. Neither Congress nor the public will be keen to answer charges that it did not prepare sufficiently for the next major threat to the security of America or its allies.

Conversely, setting technological requirements for transformation based on the speed and level of technological progress in commercial markets will make planning for NCW largely a reactive exercise, one that does not emphasize core Navy competencies. Decisions may thus be ceded to politicians, salesmen, or scientists and engineers. Although Congress may be generally sympathetic to new technologies, members are sensitive to employment levels and federal expenditures in key districts. There appears to be little need to refresh technologies at the rate maintained by some parts of the commercial sector or to worry about "Moore's Law" writ large. Given what we know about the resources available to our allies and potential allies (including their defense industrial capabilities), as well as those of adversaries and potential adversaries, if the U.S. Navy, or the military generally, adopts the commercial sector as its model, it will in the end be racing against itself, disrupting the fleet unnecessarily, and eroding public goodwill through disproportionate spending for marginal improvements.

Even if the Navy overcomes internal resistance to transformation, and even if its relationship with the defense industrial sector differs from the usual commercial customer-supplier relationship in ways that will facilitate development of the Navy after Next, network-centric warfare may still be derailed. Procurement of modern weapon systems increasingly strains the Navy's capabilities in technology acquisition. Buyers need to know what to buy, from whom to buy it, and what price to pay; technology development, however, is not a core competency of the Navy, at least not to the extent that it is of the technology-oriented firms that meet most demands for innovations. Operational requirements must be translated from "statements of objectives" into specific project plans for which the acquisition community can write contracts. The Navy needs a way to make sure that these technological requirements are attainable with reasonable investments of time and resources. For these reasons the Navy must have access to the core competencies of specialized systems integration and technology management houses.

This is especially true in that congressional politics and the Navy's comfort level with traditional contractors argue for maintaining relationships with the established defense-industrial sector. The long-standing relationships that exist between members and industry are based on a powerful confluence of money and electoral politics. If the transition to NCW threatens established sectors of the defense industrial base, Congress can and will make it politically difficult to move forward with the new programs and the cancellations necessary to achieve an NCW future; the Army's resistance to terminating the Crusader artillery system suggests just how costly political battles are for all parties—the services, DoD, Congress, and, ultimately, firms whose work may be cut. Given appropriate incentives, however, the defense industrial sector is fully capable of supporting transformation, implementing it with sustaining innovations and joint ventures that combine start-ups' disruptive innovations with established firms' customer understanding. The United States should be able to buy the Navy after Next without bankrupting the current defense firms.

When and where transformation could be viewed as a threat to the business bases of defense industrial firms, they can and will exert powerful lobbying pressure to delay or divert transformation. Any individual firms that can be persuaded to favor transformation, and certainly the defense industry as a whole, would be powerful allies in building political and budgetary support. Therefore, innovators inside the military must join forces with innovators in Congress and industry to push the transformation agenda forward.

Government and Nongovernment Interviews

Government Interviews

Army Science Board

Congressional Budget Office

Congressional Research Service

DARPA

    Information Systems Office

    Tactical Technology Office

Defence Evaluation Research Agency, U.K.

Defense Contract Management Agency

    Industrial Analysis Center

Defense Systems Management College

Department of Commerce

    Office of Strategic Industries and Economic Security, Bureau of Export Administration

Department of the Navy

    Chief Technology Officer

    Office of the Assistant Secretary of the Navy (Research, Development, & Acquisition)

Industrial College of the Armed Forces, National Defense University

NAVAIR

NAVSEA

    Innovation Center, Naval Surface Warfare Center, Carderock Division

    Shipbuilding Technologies Department, NSWC, Carderock Division

Navy Warfare Development Command

NAWCWD, China Lake

Office of Naval Research

 Industrial and Corporate Programs Office

 International Field Office, Asia

 Office of the Chief Scientist

 Office of the Executive Director

Office of the Secretary of Defense, Advanced Systems and Concepts

Office of the Under Secretary of Defense (AT&L)

 Industrial Capabilities and Assessments Directorate

 Interoperability

OpNav, N-911

SPAWAR Headquarters, San Diego

SPAWAR Systems Center San Diego

Swedish Defence Research Agency

U.S. Joint Forces Command

Nongovernment Interviews

Aerospace Corporation

ANSER

Anteon Corporation

Applied Physics Laboratory

Association for Unmanned Vehicle Systems International

BAE Systems

Belfer Center for Science and International Affairs, Kennedy School of Government, Harvard University

Boeing

 Phantom Works

 Washington Studies and Analysis

Bollinger Shipyards, Inc.

Booz Allen Hamilton

Capital Synergy Partners

Carnegie Mellon Software Engineering Institute

Cherokee Information Systems

Friede Goldman Halter, Halter Marine, Inc.

General Atomics Aeronautical Systems, Inc.

General Dynamics

    Corporate headquarters

    Bath Iron Works

    Electric Boat (Groton and Quonset Point)

    NASSCO

Hood Technology Corporation

The Insitu Group

JSA Partners, Inc.

L3 Communications, Ocean Systems

Litton Integrated Systems

Litton Ship Systems

    Avondale Industries

    Ingalls Shipbuilding

Lockheed Martin

    Corporate headquarters

    Naval Electronics and Surveillance Systems—Surface Systems

    Space Systems

Logicon

Mercury Computer

Microsoft Corporation, Government Programs

MITRE Corporation

National Defense Industrial Association

Newport News Shipyard

    Innovation Center

    Virginia Advanced Shipbuilding and Carrier Integration Center

Northrop Grumman

    Analysis Center

    Electronic Systems and Sensors Sector

        Oceanic and Naval Systems

        Sector headquarters

    Integrated Systems Sector

        Air Combat Systems

        Unmanned Systems

Raytheon

    Missile Systems

    Naval and Maritime Systems

Research, Analysis and Engineering, Inc.

SAIC

Schafer Corporation

Solipsys

SYNTEK

Todd Pacific Shipyards

List of Abbreviations

A
AAW	antiair warfare
ACTD	Advanced Concept Technology Demonstration
APL	Applied Physics Laboratory
ASW	antisubmarine warfare

C
C4ISR	command, control, communications, computer, intelligence, surveillance, and reconnaissance
CEC	Cooperative Engagement Capability
CNAN	Capabilities Navy after Next
CNO	Chief of Naval Operations
COP	common operational picture
COTS	commercial-off-the-shelf technologies
CVNX	[future aircraft carrier program]

D
DARPA	Defense Advanced Research Projects Agency
DCI	Defense Capabilities Initiative
DD(X)	[multimission surface combatant program]
DoD	Department of Defense

E
EADS	European Aeronautic Defense and Space Company
EBO	effects-based operations
EC5G	Expeditionary Command and Control, Communications, Computers, and Combat Systems Grid

	ESG	expeditionary sensor grid
F	**FAR**	federal acquisition regulations
	FFRDC	federally funded research and development center
G	**GCS**	ground control station
	GD	General Dynamics
I	**ISR**	intelligence, surveillance, and reconnaissance
	IT	information technology
	IT-21	Information Technology for the Twenty-first Century
L	**LCS**	Littoral Combat Ship
	LPD	[amphibious transport, dock]
M	**MIT**	Massachusetts Institute of Technology
N	**NASA**	National Aeronautics and Space Administration
	NASSCO	National Steel and Shipbuilding Company
	NATO	North Atlantic Treaty Organization
	NAVSEA	Naval Sea Systems Command
	NCO	network-centric operations
	NCW	network-centric warfare
	NFN	Naval Fires Network
	NMCI	Navy–Marine Corps Intranet
	NWDC	Navy Warfare Development Command (Newport, R.I.)
O	**OSD**	Office of the Secretary of Defense
	OT&E	operational test and evaluation
	OTA	Office of Technology Assessment
P	**PERT**	Program Evaluation and Review Technique
	POET	Phase One Evaluation Team

R	**R&D**	research and development
	RMA	revolution in military affairs
	RMP	Radar Modernization Program
S	**S&T**	science and technology
	SAGE	Semi-Automatic Ground Environment
	SAIC	Science Applications International Corporation
	SDI	Strategic Defense Initiative
	SDII	SDI Institute
	SEI	Software Engineering Institute
	SI	systems integration
	SOSUS	Sound Surveillance System
	SPAWAR	Space and Naval Warfare Systems Command
	SRD	Strategic Research Department (of the Naval War College Center for Naval Warfare Studies)
	STL	Space Technology Laboratory
T	**TCN**	Tactical Component Network
	TUAV	tactical unmanned aerial vehicle
U	**UARC**	university applied research center
	UAV	unmanned aerial vehicle
	UCAV	unmanned combat aerial vehicle
	UGV	unmanned ground vehicle
	USN	U.S. Navy
	USV	unmanned surface vehicle
	UUV	unmanned underwater vehicle
	UV	unmanned vehicle
W	**WeCAN**	Web Centric Anti-Submarine Warfare Net

Notes

1. Paul Bracken, "The Military after Next," *Washington Quarterly* 16, no. 4 (Autumn 1993), pp. 157–74.

2. For examples of the industry perspective, see John R. Harbison, Thomas S. Moorman, Jr., Michael W. Jones, and Jikun Kim, *U.S. Defense Industry: An Agenda for Change* (Booz Allen Hamilton, 2000); and Defense Science Board Task Force, *Preserving a Healthy and Competitive U.S. Defense Industry to Ensure our Future National Security*, Final Briefing (November 2000).

3. *Joint Vision 2020* is available at www.dtic.mil/jv2020/.

4. For the Army's vision and Army transformation, go to www.army.mil/armyvision. For a useful overview of Army transformation issues see Edward F. Bruner, *Army Transformation and Modernization: Overview and Issues for Congress*, RS20787 (Washington, D.C.: Congressional Research Service, Library of Congress, April 4, 2001); and Bruce R. Nardulli and Thomas L. McNaugher, "The Army: Toward the Objective Force," in *Transforming America's Military*, ed. Hans Binnendijk (Washington, D.C.: National Defense University, 2002), pp. 101–28.

5. The Marine Corps, on the other hand, has always recognized that it is an expeditionary force. In *Marine Corps Strategy 21* it bills itself as "the premiere 'total force in readiness.'" Headquarters, U.S. Marine Corps, *Marine Corps Strategy 21*, Washington, D.C., November 2000. Available at www.usmc.mil/templateml.nsf/25241abbb036b230852569c4004eff0e/$FILE/strategy.pdf.

6. The U.S. Air Force's *Vision 2020* can be found at www.af.mil/vision/.

7. From *The Aerospace Force: Defending America in the 21st Century*, p. iii, at www.af.mil/lib/taf.pdf. Overviews of Air Force transformation issues are provided by Christopher Bolkom, *Air Force Transformation and Modernization: Overview and Issues for Congress*, RS20787 (Washington, D.C.: Congressional Research Service, Library of Congress, June 1, 2001); and David Ochmanek, "The Air Force: The Next Round," in Binnendijk, ed., *Transforming America's Military*, pp. 159–90.

8. Navy Warfare Development Command [hereafter NWDC], *Network Centric Operations: A Capstone Concept for Naval Operations in the Information Age* (Newport, R.I.:, draft dated June 19, 2001), p. 1. The growing literature on NCW includes David S. Alberts, John J. Garstka, and Frederick P. Stein, *Network Centric Warfare: Developing and Leveraging Information Superiority*, 2nd ed. (Washington, D.C.: C4ISR Cooperative Research Program, 1999); Vice Admiral Arthur K. Cebrowski and John J. Garstka, "Network-Centric Warfare: Its Origin and Future," U.S. Naval Institute *Proceedings*, January 1998, pp. 28–35; Committee on Network-Centric Naval Forces, Naval Studies Board, *Network-Centric Naval Forces: A Transition Strategy for Enhancing Operational Capabilities* (Washington, D.C.: National Academy Press, 2000), William D. O'Neil, "The Naval Services: Network-Centric Warfare," in Binnendijk, ed., *Transforming America's Military*, pp. 129–58; and Edward P. Smith, "Network-Centric Warfare: What's the Point?" *Naval War College Review* 54, no. 1 (Winter 2001), pp. 59–75.

9. Robert Little, "Bush Makes Defense Firms Nervous," *Baltimore Sun*, January 21, 2001; Erin E. Arvedlund, "Starship Troopers: New Weaponry Will Shake Up the Defense Industry—and Investors," *Barron's*, February 12, 2001; Gopal Ratnam and Jason Sherman, "High-Stakes Gamble," *Defense News*, May 28–June 3, 2001, p. 22; "Transformed: A Survey of the Defence Industry," *Economist*, July 20, 2002, pp. 1–16; Craig Covault, "Net-Centric Ops, UAVs Reshape Battlefields and Boardrooms," *Aviation Week and Space Technology*, July 22, 2002, p. 163, and Vago Muradian, "Questions, but No Answers: Industry Chiefs See Future Rife with Uncertainty," *Defense News*, July 29–August 4, 2002, pp. 1 and 9.

10. Michael O'Hanlon, *Technological Change and the Future of Warfare* (Washington, D.C.: Brookings Institution, 2000).

11. Even if definitive answers cannot be provided at this stage of the transformation process.

12. For serious treatments of the forces at work see Daniel Bell, *The Coming of Post-Industrial Society: A Venture in Social Forecasting* (New York: Basic Books, 1999), and Manuel Castells, *The Rise of the Network Society*, 2nd ed. (Oxford: Blackwell, 2000). For popular treatments, see Thomas L. Friedman, *The Lexus and the Olive Tree: Understanding Globalization* (New York: Farrar, Straus, and Giroux, 1999) and James Gleick, *Faster: The Acceleration of Just About Everything* (New York: Pantheon, 1999). Works that have received far more attention in military circles than they deserve include Kevin Kelly, *New Rules for the New Economy: 10 Radical Strategies for a Connected World* (New York: Viking, 1998); Alvin Toffler, *The Third Wave* (New York: William Morrow, 1980) and *Powershift: Knowledge, Wealth, and Violence at the Edge of the 21st Century* (New York: Bantam Books, 1990); and Alvin and Heidi Toffler, *Creating a New Civilization: The Politics of the Third Wave* (Atlanta: Turner, 1995).

13. See John Arquilla and David Ronfeldt, eds., *In Athena's Camp: Preparing for Conflict in the Information Age* (Santa Monica, Calif.: RAND, 1997); Eliot A. Cohen, "A Revolution in Warfare," *Foreign Affairs* 75, no. 2 (March–April, 1996), pp. 37–54; Victor A. DeMarines, with David Lehman and John Quilty, "Exploiting the Internet Revolution," in *Keeping the Edge: Managing Defense for the Future*, ed. Ashton B. Carter and John P. White (Cambridge, Mass., and Stanford, Calif.: Preventive Defense Project, 2000), pp. 61–102; Joseph S. Nye, Jr., and William A. Owens, "America's Information Edge," *Foreign Affairs* 75, no. 2 (March–April 1996), pp. 20–36; and Bill Owens, with Ed Offley, *Lifting the Fog of War* (New York: Farrar, Straus and Giroux, 2000). For a more popular account see Alvin and Heidi Toffler, *War and Anti-War: Making Sense of Today's Global Chaos* (New York: Warner, 1993).

14. See, for example, Cebrowski and Garstka, "Network-Centric Warfare: Its Origin and Future," Alberts, Garstka, and Stein, *Network Centric Warfare: Developing and Leveraging Information Superiority*, especially pp. 15–23. In his "Preface" to the April 2000 *Navy Planning Guidance*, Admiral Jay L. Johnson, then Chief of Naval Operations, wrote of building "a Navy for the Information Age" and its "transformation to a network-centric and knowledge-superior force"; Chief of Naval Operations, *Navy Planning Guidance: With Long Range Planning Objectives* (Washington, D.C.: Department of the Navy, April 2000), p. i. According to the *Navy Planning Guidance*, p. 51, "The Navy of the future will conduct all operations based on the concept of *Network Centric Operations (NCO)*" (emphasis in the original). Similarly, the then Vice Chief of Naval Operations declared that "we are moving away from a platform-centered Navy to one being built around data networks. . . . [O]ur concept of operations will use as its basis an integrated, common network"; William J. Fallon, "Fighting to Win in the Littoral and Beyond," *Armed Forces Journal International*, June 2001, pp. 67 and 68. Vice Admiral Dennis McGinn has asserted, "Investment in networks and sensors is transformational"; quoted in Robert Holzer, "U.S. Navy Mulls Fundamental Shift in Tactics, Funds," *Defense News*, May 7, 2001, p. 1. A useful discussion of NCW/NCO is provided by Scott C. Truver, "Tomorrow's U.S. Fleet," U.S. Naval Institute *Proceedings*, March 2001, pp. 102–10. For a comparison of U.S. and Swedish versions of NCW see Nick Cook, "Network-Centric Warfare: The New Face of C4I," *Interavia*, February 2001, pp. 37–39. Cautionary notes are provided by Thomas P. M. Barnett, "The Seven Deadly Sins of Network-Centric Warfare," U.S. Naval Institute *Proceedings*, January 1999, pp. 36–39; Richard J. Harknett and the JCISS Study Group, "The Risks of a Networked Military," *Orbis* 44, no. 1 (Winter 2000), pp. 127–43; Franklin Spinney, "What Revolution in Military Affairs?" *Defense Week*, April 23, 2001.

15. Not all nodes, of course, are created equal. Some are more complex and therefore more expensive than others. The point is that

networked nodes should be simpler and lower in cost than stand-alone nodes.

16. As of this writing, there was no "official" Navy document or statement that describes NCW. Indeed, there is no real consensus among its proponents as to precisely what NCW is or entails. Its proponents charitably view NCW as a dynamic, living, evolving concept. Skeptics are more inclined to characterize NCW as a moving target riddled with ambiguities and informed by dubious analogies. In a definition attributed to John Garstka, NCW is "warfare which harnesses information technologies in the form of global sensor, connectivity, and engagement grids to achieve a common operational picture that will lead to self-synchronization, massed effects, and the desired lock-out of a given enemy's courses of action"; see Robert Odell, Bruce Wald, Lyntis Beard, with Jack Batzler and Michael Loescher, *Taking Forward the Navy's Network Centric Warfare Concept: Final Report*, CRM 99-42.10 (Alexandria, Va.: Center for Naval Analyses, May, 1999), p. 11. The Naval Studies Board's Committee on Network-Centric Naval Forces defined network-centric operations as "military operations that exploit state-of-the-art information and networking technology to integrate widely dispersed human decision makers, situational and targeting sensors, and forces and weapons into a highly adaptive, comprehensive system to achieve unprecedented mission effectiveness"; Committee on Network-Centric Naval Forces, Naval Studies Board, *Network-Centric Naval Forces*, p. 12. The Navy Warfare Development Command describes NCO as "deriving power from the rapid and robust networking of well-informed, geographically dispersed warfighters. They create overpowering tempo and a precise, agile style of maneuver warfare"; NWDC, *Network Centric Operations: A Capstone Concept for Naval Operations in the Information Age* (Newport, R.I.: Navy Warfare Development Command, draft dated June 19, 2001), p. 1. Available at www.nwdc.navy.mil/Concepts/capstone_concept.asp.

17. NWDC, *Network Centric Operations: A Capstone Concept for Naval Operations in the Information Age*.

18. On information and knowledge advantage see www.nwdc.navy.mil/Concepts/IKA.asp.

19. On effects-based operations see www.nwdc.navy.mil/Concepts/EBO.asp.

20. Truver, "Tomorrow's U.S. Fleet," p. 103.

21. NWDC, *Network Centric Operations*, p. 9.

22. Ibid., p. 11.

23. Ibid., p. 10.

24. That NCW is no longer merely a service vision is illustrated by the DoD report to Congress on NCW: Department of Defense, *Network Centric Warfare* (Washington, D.C.: Department of Defense, 27 July 2001); available at www.c3i.osd.mil/NCW/. This report reviews the NCW visions of all of the services. The joint aspects of NCW are also highlighted in John J. Garstka, "Network Centric Warfare: An Overview of Emerging Theory," available at www.mors.org/Pubs/phalanx/dec00/feature.htm; John G. Roos, "An All-Encompassing Grid," *Armed Forces Journal International*, January 2001, pp. 26–35; Hunter Keeter, "Cebrowski: Joint Philosophy Fosters Network Centric Warfare," *Defense Daily*, April 12, 2002, p. 8; and Fred P. Stein, "Observations on the Emergence of Network Centric Warfare," available at www.dodccrp.org/steinncw.htm and as "Information Paper: Observations on the Emergence of Network-Centric Warfare" at www.dtic.mil/jcs/j6/education/warfare.html. Emerging congressional support for NCW as a joint vision is indicated by Joseph Lieberman, "The Future Is Networked," *Defense News*, August 21, 2000, p. 15.

25. On assured access see www.nwdc.navy.mil/Concepts/AA.asp.

26. NWDC, *Network Centric Operations*, p. 10.

27. On forward sea-based forces see www.nwdc.navy.mil/Concepts/FSBF.asp.

28. NWDC, *Network Centric Operations*, pp. 4–5.

29. Ibid., p. 6.

30. For background on CEC, IT-21, and NMCI, see Ronald O'Rourke, *Navy Network-Centric Warfare Concept: Key Programs and Issues for Congress*, RS20557 (Washington, D.C.: Congressional Research Service, Library of Congress, June 6, 2001). On IT-21 see J. Cutler Dawson, Jr., James M. Fordice, and Gregory M. Harris, "The IT-21 Advantage," U.S. Naval Institute *Proceedings*, December 1999, pp. 28–32. To Admiral Vernon E. Clark, the Chief of Naval Operations, the NMCI is "the gateway to transformation"; see Department of the Navy, *Electronic Business Strategic Plan 2001–2002*, available at www.ec.navsup.navy.mil_eb/strategic_plan_toc.asp.

31. NWDC, *Expeditionary Sensor Grid*, undated brief, p. 4. See also Robert Holzer, "Massive Sensor Grid May Reshape U.S. Navy Tactics," *Defense News*, May 14, 2001, pp. 1 and 4; and Catherine MacRae, "Services, DARPA Doing Early Research on 'Expeditionary Sensor Grid,'" *Inside the Pentagon*, June 21, 2001.

32. NWDC, *The Expeditionary Sensor Grid: Gaining Real-Time Battlespace Awareness in Support of Information and Knowledge Advantage*, post-workshop draft, June 19, 2001, p. 3.

33. *Naval Transformation Roadmap: Power and Access . . . from the Sea* (Washington, D.C.: Department of the Navy, 2002). For reports on FORCEnet see Gopal Ratnam, "New Office to Drive U.S. Navy Transformation," *Defense News*, April 8–14, 2002, p. 6; and Gail Kaufman and Gopal Ratnam, "U.S. Navy Releases Broad Transformation Outline," *Defense News*, April 15–21, 2002, p. 8.

34. This set of capabilities is to be developed in a phased process. During the first phase, 2002–2004, the focus will be on improving networks, sensors, people, and weapons, with networks and sensors the highest priorities. People and infrastructure will be accorded highest priority during the second stage, 2004–2010, and platform and infrastructure improvements are to be added to the agenda. Platform and infrastructure improvements join the list of high-priority efforts during the third stage, 2010–2020. On SEA POWER 21, see Admiral Vernon Clark, "SEA POWER 21: Operational Concepts for a New Era," remarks delivered at the Current Strategy Forum, Naval War College, Newport, Rhode Island, 12 June 2002.

35. A FORCEnet office, directed by Vice Admiral Dennis McGinn, USN, Deputy Chief of Naval Operations (N6/N7 Warfare Requirements and Programs), has been established in OpNav; in July 2002 the Naval Network Warfare Command (NETWARCOM) was stood up at the Little Creek Naval Amphibious Base.

36. It should be noted that the Army relies on sealift for its deployments and thus has an inherent interest in ship concepts and technologies that could transport its forces to future battlefields more quickly and efficiently than those available today.

37. On the concept of a "system of systems," see William A. Owens, "The Emerging System of Systems," U.S. Naval Institute *Proceedings*, May 1995, pp. 36–39; and William A. Owens, "The Emerging U.S. System-of-Systems," Strategic Forum 63 (Washington, D.C.: Institute for National Strategic Studies, National Defense University, February 1996).

38. As quoted in Christopher J. Castelli, "Northrop Executive: Technology Integration Will Be a Tough Task," *Inside the Navy*, April 30, 2001.

39. As quoted in MacRae, "Services, DARPA Doing Early Research on 'Expeditionary Sensor Grid.'"

40. Opening Statement of Under Secretary of Defense (AT&L) E. C. "Pete" Aldridge, Jr., before the American Institute of Astronautics and Aeronautics, Washington, D.C., February 19, 2002.

41. Author interview, February 2001.

42. Author interview, May 2001.

43. Author interview, June 2001.

44. Author interview, February 2001.

45. Author interview, June 2001. Phil Condit, the chairman and CEO of Boeing, in "Industry Challenges to Achieving Government Vision," an address before the 2002 AIAA Defense Excellence Conference in Washington, D.C., on February 20, 2002, stated that "we

need to rethink our industry vision[,] ... because defense will be much different as ... transformation takes hold." Available at www.boeing.com/news/speeches/2002/condit_020220.html.

46. Author interview, November 2000.

47. For a prominent discussion of firms' efforts to manage innovation and develop salable products, see James M. Utterback, *Mastering the Dynamics of Innovation* (Boston: Harvard Business School Press, 1994).

48. Clayton M. Christensen, *The Innovator's Dilemma: When New Technologies Cause Great Firms to Fail* (Boston: Harvard Business School Press, 1997). See also Joseph L. Bower and Clayton M. Christensen, "Disruptive Technologies: Catching the Wave," *Harvard Business Review*, January–February 1995, pp. 43–53; Clayton M. Christensen and Richard S. Tedlow, "Patterns of Disruption in Retailing," *Harvard Business Review*, January–February 2000, pp. 42–45; Clayton M. Christensen and Michael Overdorf, "Meeting the Challenge of Disruptive Change," *Harvard Business Review*, March–April 2000, pp. 67–76; Clayton M. Christensen, Mark V. Johnson, and Darrell K. Rigby, "Foundations for Growth: How to Identify and Build Disruptive New Businesses," *MIT Sloan Management Review* 43, no. 3 (Spring 2002), pp. 22–31; and Clayton M. Christensen, "The Rules of Innovation," *Technology Review* (June 2002), pp. 33–38. For applications to defense planning, see Capt. Terry C. Pierce, "Jointness Is Killing Naval Innovation," U.S. Naval Institute *Proceedings* (October 2001), pp. 68–71; and Fred E. Saalfeld and John F. Petrik, "Disruptive Technologies: A Concept for Moving Innovative Military Technologies Rapidly to Warfighters," *Armed Forces Journal International* (May 2001), pp. 48–52.

49. Christensen, *The Innovator's Dilemma*, p. xv.

50. The literature on innovation has stressed for some time the importance of the relationships between technology firms and their customers. The leading study of these relationships is by Eric von Hippel, *The Sources of Innovation* (New York: Oxford University Press, 1988).

51. Christensen, *The Innovator's Dilemma*, p. 189.

52. Harvey M. Sapolsky, "On the Theory of Military Innovation," *Breakthroughs* 9, no. 1 (Spring 2000), p. 39.

53. Leading supply firms are reluctant to pursue disruptive innovations for an additional reason: the initially small, down-market demand is too small to yield the revenue and profit growth that established firms seek, especially when the disruptive market's prospects are compared to the existing sales base of the successful firm. For a concise summary of this reasoning, see Clayton Christensen, Thomas Craig, and Stuart Hart, "The Great Disruption," *Foreign Affairs* 80, no. 2 (March–April 2001), pp. 82–83. This reason is unlikely to apply in the context of military transformation, because the customers (military services) promise to stop buying old products, eliminating the established sales base, and to expand purchases of new equipment rapidly. Thus far, however, there is little reason to take such a promise seriously.

54. This pattern was followed with the widespread introduction of missiles into the military arsenal—a disruptive innovation in the defense industry of the 1960s. See G. R. Simonson, "Missiles and Creative Destruction in the American Aircraft Industry, 1956–1961," in *The History of the American Aircraft Industry: An Anthology*, ed. G. R. Simonson (Cambridge, Mass.: MIT Press, 1968), pp. 230, 241.

55. For an insightful and engaging discussion, see Carl H. Builder, *The Masks of War: American Military Styles in Strategy and Analysis* (Baltimore: Johns Hopkins University Press, 1989).

56. The contemporary veneration for capabilities-based planning is evident in transformation proposals.

57. There are, of course, other theories of military innovation. For a cultural interpretation, see Elizabeth Kier, *Imagining War: French and British Military Doctrine between the Wars* (Princeton, N.J.: Princeton University Press, 1997). An approach that draws

upon the domestic structures and organizational behavior literatures is provided by Matthew Evangelista, *Innovation and the Arms Race: How the United States and the Soviet Union Develop New Military Technologies* (Ithaca, N.Y.: Cornell University Press, 1988). A review of the literature on military technological dynamics is provided by Andrew L. Ross, "The Dynamics of Military Technology," in *Building a New Global Order: Emerging Trends in International Security*, ed. David Dewitt, David Haglund, and John Kirton (Toronto: Oxford University Press, 1993), pp. 106–40.

58. Barry R. Posen, *The Sources of Military Doctrine: France, Britain, and Germany between the Wars* (Ithaca, N.Y.: Cornell University Press, 1984). According to Posen, failure of the civilians to intervene often leads to political-military disintegration, with potentially disastrous consequences in times of crisis or war.

59. Peter J. Boyer, "A Different War," *New Yorker*, July 1, 2002, pp. 54–67; Frank Hoffman, "Goldwater-Nichols after a Decade," in *The Emerging Strategic Environment: Challenges of the Twenty-first Century*, ed. Willamson Murray (Westport, Conn.: Praeger, 1999), pp. 156–82; Sharon Weiner, "The Politics of Resource Allocation in the Post–Cold War Pentagon," *Security Studies* 5, no. 4 (Summer 1996), pp. 125–42.

60. Harvey M. Sapolsky and Eugene Gholz, "The Defense Industry's New Cycle," *Regulation* 24, no. 3 (Winter 2001–2002), pp. 44–49.

61. Stephen Peter Rosen, *Winning the Next War: Innovation and the Modern Military* (Ithaca, N.Y.: Cornell University Press, 1991).

62. Owen R. Cote, Jr., *The Politics of Innovative Military Doctrine: The U.S. Navy and Fleet Ballistic Missiles* (Ph.D. dissertation, Massachusetts Institute of Technology, 1995). Cote describes a defense posture featuring maritime capabilities in his "Buying ' . . . From the Sea': A Defense Budget for a Maritime Strategy," in *Holding the Line: U.S. Defense Alternatives for the Early 21st Century*, ed. Cindy Williams (Cambridge, Mass.: MIT Press, 2001), pp. 141–79.

63. Each community can even gain status by taking a leadership role in promoting a particular joint solution, although there may obviously be a natural bias in favor of service-dominated concepts.

64. Although in practice some of the "open" debate may be open only within the classified community. "Open" in this context means available for critique and improvement by other military communities.

65. Eugene Gholz and Harvey Sapolsky, "Restructuring the U.S. Defense Industry," *International Security* 24, no. 3 (Winter 1999–2000), p. 50.

66. For an intriguing discussion of the prediction process, see John H. Holland, "What Is to Come and How to Predict It," in *The Next Fifty Years: Science in the First Half of the Twenty-first Century*, ed. John Brockman (New York: Vintage Books, 2002), pp. 170–82.

67. Of course, the post–Cold War era was hardly the first period of major defense industry consolidation. For an overview of the history of the U.S. defense industry, see Aaron L. Friedberg, *In the Shadow of the Garrison State: America's Anti-Statism and Its Cold War Strategy* (Princeton, N.J.: Princeton University Press, 2000), especially, chaps. 6–7; Paul A. C. Koistinen, *Beating Plowshares into Swords: The Political Economy of American Warfare, 1606–1865* (Lawrence: University Press of Kansas, 1996); Paul A. C. Koistinen, *Mobilizing for Modern War: The Political Economy of American Warfare, 1865–1919* (Lawrence: University Press of Kansas, 1997); and Paul A. C. Koistinen, *Planning War, Pursuing Peace: The Political Economy of American Warfare, 1920–1939* (Lawrence: University Press of Kansas, 1998).

68. Remarks of Jeffrey P. Bialos, Deputy Under Secretary of Defense for Industrial Affairs, Credit Suisse First Boston Aerospace Finance Executive Symposium, April 27, 2000.

69. See Gholz and Sapolsky, "Restructuring the U.S. Defense Industry"; and Gopal Ratnam, "Industry Consolidates, but Factories Stay Open," *Defense News*, January 29, 2001, pp. 3 and 19.

70. In a sensible recognition of the efficiency disadvantages of this production-sharing arrangement in the face of substantial economies of scale in defense production, the Navy has recently negotiated an arrangement with Bath Iron Works and Northrop Grumman's Avondale Shipyard by which production of DDG 51s will be concentrated at Bath and LPD 17s at Avondale. Both shipyards will continue with modest levels of profitable production, but costs will be reduced for both programs. Andy Pasztor, "Navy Realigns Shipbuilding Jobs of Northrop, General Dynamics," *Wall Street Journal*, June 18, 2002.

71. Linda R. Cohen and Roger G. Noll, "Government Support for R&D," in *The Technology Pork Barrel*, ed. Linda R. Cohen and Roger G. Noll (Washington, D.C.: Brookings Institution, 1991), p. 25.

72. Harvey Sapolsky and Eugene Gholz, "Eliminating Excess Defense Production," *Issues in Science and Technology* 13, no. 2 (Winter 1996–97), pp. 65–70. On-line version available at www.nap.edu/issues/13.2/sapols.htm.

73. See, for instance, Ann Markusen, "The Rise of World Weapons," *Foreign Policy*, no. 114 (Spring 1999), pp. 40–51.

74. Glenn R. Simpson, "Dutch Government Expresses Ire over U.S. Threat to Block Merger," *Wall Street Journal*, April 27, 2001.

75. Mattias Axelson and Andrew James, *The Defence Industry and Globalisation: Challenging Traditional Structures*, FOA-R-00-01698-170-SE (Stockholm: Defense Research Establishment, December 2000), p. 35.

76. Eugene Gholz, "The Irrelevance of International Defense Mergers," *Breakthroughs* 9, no. 1 (Spring, 2000), pp. 3–11.

77. See Andrew L. Ross, "Defense Industry Globalization: Contrarian Observations," in *Defense Industry Globalization* (Washington, D.C.: Atlantic Council of the United States, February 2002), pp. 35–42; Peter J. Dombrowski, "The Globalization of the Defense Sector? Naval Industrial Cases and Issues," in *Globalization and Maritime Power*, ed. Sam J. Tangredi (Washington, D.C.:

National Defense University Press, December 2002).

78. John A. Alic, L. M. Branscomb, A. B. Carter, and G. L. Epstein, *Beyond Spinoff: Military and Commercial Technologies in a Changing World* (Boston: Harvard Business School Press, 1992).

79. According to former Under Secretary of Defense Jacques Gansler, "DoD must learn to capture commercial technology (both product and process technologies). . . . Perhaps most essential for the transformation of our defense acquisition practices and industrial structures is the need to bring about far greater civilian/military industrial integration." Remarks by the Hon. Jacques S. Gansler, Under Secretary of Defense (Acquisition and Technology), Tech Trends 2000 Conference, Philadelphia, April 7, 1999. Gansler made the case for what he labeled "civil/military integration" in *Defense Conversion: Transforming the Arsenal of Democracy* (Cambridge, Mass.: MIT Press, 1995).

80. For an examination of post–Cold War conversion efforts, see Greg Bischak, *US Conversion after the Cold War, 1990–1997: Lessons for Forging a New Conversion Policy*, Brief 9 (Bonn: Bonn International Center for Conversion, July 1997). A more positive assessment of the defense industry's diversification and conversion efforts is provided by Michael Oden, "Cashing In, Cashing Out, and Converting: Restructuring of the Defense Industrial Base in the 1990s," in *Arming the Future*, ed. Markusen and Costigan (New York: Council on Foreign Relations Press, 1999), pp. 74–105.

81. One future core competency of the defense industry might be the ability to serve as an intermediary between the wider commercial world and the specialized world of government procurement. This insight emerged from discussions with Martin Lundmark.

82. In the end, it is largely this reliance on ships that distinguishes the Navy from the other U.S. services.

83. See Rear Admiral Charles Hamilton, USN, and Rear Admiral Donald Loren, USN, "It's

All in the Family," U.S. Naval Institute *Proceedings*, August 2002, pp. 68–70.

84. Vice Admiral Michael Mullen, USN, Deputy Chief of Naval Operations for Resources, Requirements, and Assessments (N8), as quoted in Hunter Keeter, "Navy Six Months from Refining Industry Roles in LCS Concept," *Defense Daily*, July 18, 2002, p. 6.

85. See also Christian Bohmfalk, "Navy Sees Littoral Ship Combating Mines, Diesel Subs, Small Boats," *Inside the Navy*, January 28, 2002, pp. 5–6.

86. Jonathan Rauch, "The New Old Economy: Oil, Computers, and the Reinvention of the Earth," *Atlantic Monthly*, January 2001, pp. 35–49.

87. Author interview, November 2000.

88. For more extended discussions of shipyard ownership consolidation, see three studies by Ronald O'Rourke, all published by the Congressional Research Service, Library of Congress: *Navy Major Shipbuilder Ownership Consolidation: Issues for Congress*, RL3051 (Washington, D.C.: July 7, 1999); *Navy Shipbuilding: Proposed Mergers Involving Newport News Shipbuilding—Issues for Congress*, RL30969 (Washington, D.C.: May 22, 2001); and *Navy Shipbuilding: Recent Shipyard Mergers—Background and Issues for Congress*, RL31400 (Washington, D.C.: May 3, 2002).

89. On the overall health of the American shipbuilding industry, especially as it relates to national security, see U.S. Department of Commerce, *National Security Assessment of the U.S. Shipbuilding and Repair Industry* (Washington, D.C.: U.S. Department of Commerce, Bureau of Export Administration, Office of Strategic Industries and Economic Security, May 2001).

90. On CVNX, see Ronald O'Rourke, *Navy CVNX Aircraft Carrier Program: Background and Issues for Congress*, RS20643 (Washington, D.C.: Congressional Research Service, Library of Congress, May 23, 2002). Northrop Grumman, which has sought to position itself as an RMA firm, has expressed a preference for modifying Ingalls's *Wasp*-class amphibious ships rather than developing a new design for the LHA replacement (LHA-R). See Christopher J. Castelli, "Northrop Exec: Repeating Existing LHD Design Is Most Cost Effective," *Inside the Navy*, April 1, 2002, p. 12.

91. Thomas C. Hone, "Force Planning Cycles: The Modern Navy as an Illustrative Case of a Frustrating Trend," *Defense Analysis* 9, no. 1 (April 1993), pp. 31–42.

92. Harvey M. Sapolsky, "Equipping the Armed Forces," in *National Security and the U.S. Constitution*, eds. George Edwards and W. Earl Walker (Baltimore: Johns Hopkins University Press, 1988), pp. 121–35.

93. This tendency toward complexity may also apply to NCW-friendly platforms (LCSs and UAVs), for which simplicity and affordability are key points used in justification. The interaction between political and technological uncertainty may be one limit on the ability of NCW advocates to get their vision adopted by the acquisition community.

94. Geoffrey Wood, "The Rise of Unconventional Naval Platforms," *Military Technology*, May 2002, pp. 58–63.

95. On stealth see James A. King, "Stealth Means Survivability," U.S. Naval Institute *Proceedings* (December 2001), pp. 80–83.

96. Norman Friedman, *Seapower and Space: From the Dawn of the Missile Age to Net-centric Warfare* (Annapolis, Md.: Naval Institute Press, 2000).

97. Keeter, "Navy Six Months from Refining Industry Roles in LCS Concept," p. 6.

98. Gholz and Sapolsky, "Restructuring the U.S. Defense Industry."

99. Randy Woods, "Navy Briefing Estimates Littoral Ships Could Cost $542 Million Each," *Inside the Navy*, July 22, 2002, p. 1.

100. Author interviews, February 2001 and March 2002.

101. Some advocates of network-centric warfare emphasize that their vision would require the elimination of the battle group and the amphibious ready group as force-sizing metrics for the Navy. All ships in the fleet would

cooperate with each other through the network.

102. The benefits of doing so may be reduced by time and distance, however.

103. Of course, complex communication systems that are particularly robust, redundant, secure, etc.—all traditional military network-performance metrics—may require complex naval architecture to install, especially if stealth remains an important ship-performance metric.

104. Numerous important questions about modularity remain. Where, for example, will the "plug and play" exchange take place—in-theater or in the continental United States? At a depot or private shipyard? How long will the exchange require—days or weeks (in which case it could influence particular crises or specific missions), or months (in which case it could respond, at best, to long-term ebbs and flows of international politics)?

105. Of course, commercial vessels normally operate on a "point-to-point" basis and are not designed or intended to survive when severely damaged.

106. During author interviews with representatives of most of the Big Six, engineers and strategic planners seemed eager to demonstrate how well their current programs fit transformation and NWC requirements. They also asked numerous variants of the "So, what does NCW really mean?" question. In short, they had discovered the importance of the vision and wanted to understand what its adoption would mean for their own businesses. By contrast, smaller and commercial yards demonstrated much less knowledge of transformation. Their executives listened, sometimes only politely, to our descriptions of NCW and naval transformation.

107. Halter Marine was acquired by Vision Technologies Kinetics, Inc., a subsidiary of Singapore Technologies Engineering, Ltd., from Friede Goldman Halter, Inc., in July 2002. It is now known as VT–Halter Marine Group.

108. Author interview, February 2001.

109. On the *Joint Venture*, see Admiral Robert J. Natter, U.S. Navy, "Meeting the Need for Speed," U.S. Naval Institute *Proceedings*, June 2002, pp. 65–67; and Harold Kennedy, "U.S. Services Test Aussie-Built Catamaran," *National Defense*, April 2002, pp. 30–31.

110. Author interview, March 2002. On Bollinger, see Gopal Ratnam, "Small Ships, Big Opportunities," *Defense News*, May 27–June 2, 2002, p. 14.

111. However, one co-author has specifically argued that naval shipbuilding has not been globalized. See, Dombrowski, "The Globalization of the Defense Sector? Naval Cases and Issues."

112. For discussion in the context of a specific (failed) program that was widely perceived as innovative in many of the senses of network-centric warfare, see Robert S. Leonard, Jeffrey A. Drezner, and Geoffrey Sommer, *The Arsenal Ship: Acquisition Process Experience* (Santa Monica, Calif.: RAND, 1999).

113. Ronald O'Rourke, *Navy DD(X) Future Surface Combatant Program: Background and Issues for Congress*, RS21059 (Washington, D.C.: Congressional Research Service, Library of Congress, May 10, 2002); Ronald O'Rourke, *Coast Guard Deepwater Program: Background and Issues for Congress*, RS21019 (Washington, D.C.: Congressional Research Service, Library of Congress, May 23, 2002).

114. Unmanned vehicles include unmanned ground vehicles (UGVs), unmanned surface vehicles (USVs), unmanned underwater vehicles (UUVs), and unmanned aerial vehicles (UAVs). Unmanned combat aerial vehicles (UCAVs) are a subcategory of UAVs.

115. Some smaller segment of missions may be entirely new, because they are not currently part of the existing operational lexicon.

116. An informative discussion of the types of UAVs, their advantages and disadvantages, and the roles they might play is provided by David B. Glade II, "Unmanned Aerial Vehicles," in *The Technological Arsenal: Emerging Defense Capabilities,* ed. William C. Martel

(Washington, D.C.: Smithsonian Institution Press, 2001), pp. 173–95.

117. This judgment is subject to dispute. Some engineers in the UAV business argue that current-generation UAVs represent an evolutionary step—either from manned to unmanned vehicles or from cruise missiles to unmanned vehicles. Under these interpretations, the UAV industry's antecedents lie with the aerospace and missile industries, respectively. Others suggest that antecedents of the UAV business extend all the way back to World War I, when target drones were first built. For our purposes, history matters less than the simple fact that relatively large-scale production of UAVs did not begin until the late 1980s.

118. Ryan Aeronautical was recently acquired by Northrop Grumman, but it maintains a separate identity within the larger defense conglomerate.

119. Eric Labs, *Options for Enhancing the Department of Defense's Aerial Vehicles Programs* (Washington, D.C.: Congressional Budget Office, September 1998), table 1.

120. John Birkler, Giles Smith, Glenn A. Kent, and Robert V. Johnson, *An Acquisition Strategy, Process, and Organization for Innovative Systems* (Santa Monica, Calif.: RAND, 2000), pp. 8–9.

121. TRW was acquired by Northrop Grumman in the early summer of 2002.

122. Data on European and Israeli UAV firms derived from Kenneth Munson, ed., *Jane's Unmanned Aerial Vehicles and Targets*, no. 17 (Coulsdon, Surrey, U.K., and Alexandria, Va.: Jane's Information Group, Ltd., December 2001).

123. Author interview, June 2002.

124. Project on Government Oversight, "Fighting with Failure Series: Case Studies of How the Pentagon Buys Weapons, Predator Unmanned Aerial Vehicle," March 22, 2002, available at www.pogo.org/mici/failures/predator.htm. For a different take on the numbers lost, see Ron Laurenzo, "Combat Losses Account for Most Predators," *Defense Week*, May 28, 2002, p. 2.

125. Amy Butler, "Air Force to Propose $750 Million Cut to Global Hawk UAV in POM," *Inside the Air Force*, July 12, 2002, p. 1. See also Robert Wall, "Costs Spur Drive to Tweak Global Hawk," *Aviation Week and Space Technology*, June 17, 2002, p. 28. How this figure was derived and how accurate it is are less important at this stage than that low-cost, potentially disposable UAVs show signs of becoming more expensive and less disposable.

126. Author interview, August 2001.

127. William M. Arkin, "Unmanned Planes Face Threats from Near and Far," *Los Angeles Times*, February 3, 2002; and David A. Fulghum, "Unmanned Designs Expand Missions and Lower Costs," *Aviation Week and Space Technology*, July 29, 2002, p. 28.

128. For a discussion of this issue in the context of the Army's tactical UAV (TUAV) program, see Glenn W. Goodman, Jr., "Manned-Unmanned Synergy: US Army's UAV-Related Efforts Gain Momentum," *Armed Forces Journal International*, July 2002, pp. 56–61.

129. The possibility of datalink interruption temporarily blinding a UAV pilot and causing air traffic control problems is better thought of as part of the "datalink quality" performance metric, although it obviously has a potential impact on situational awareness as well.

130. Author interview, June 2002.

131. "U.S. Grounds Hawk Spy Plane," *Washington Post*, July 11, 2002, p. 11.

132. Author interview, May 2002. See also Clint Housh, "UAV/UCAV," Naval Aviation Systems Team, Naval Air Warfare Center, January 19, 2001.

133. Author interview, June 2002.

134. Northrop Grumman is also working on small contracts for trade studies and risk reduction as part of the DARPA/Navy UCAV program. Its related work on the Pegasus UAV, while not directly funded by the Navy, might help identify performance metrics in the same way that Boeing's X-45 could, but the Northrop Grumman effort as currently funded is less likely to build an important customer-supplier relationship for UCAVs.

135. Thomas P. Ehrhard, *Unmanned Aerial Vehicles in the United States Armed Services: A Comparative Study of Weapon System Innovation* (Ph.D. dissertation, Johns Hopkins University, 2001).

136. Amy Svitak, "Disjointed First Steps: U.S. Services' Transformation Plans Compete, Don't Cooperate," *Defense News*, August 19–25, 2002, p. 1.

137. Other prime contractors perform a similar, product-specific kind of system integration for sensor equipment, propulsion equipment, and other major platform components.

138. Scott Tumpak, "Limit Super Primes," *Defense News*, July 15–21, 2002, p. 23; Andrew Chuter, "Honeywell Eyes FCS Systems Integration," *Defense News*, July 29–August 4, 2002, p. 4.

139. Harvey M. Sapolsky, Eugene Gholz, and Allen Kaufman, "Security Lessons from the Cold War," *Foreign Affairs* 78, no. 4 (July–August, 1999), pp. 77–89.

140. For a related discussion of the tensions between operational Navy commanders and research scientists at the Office of Naval Research, see Harvey M. Sapolsky, *Science and the Navy: The History of the Office of Naval Research* (Princeton, N.J.: Princeton University Press, 1990), pp. 86, 89, 96–98.

141. The defense business remains a political one, and it is unrealistic to believe that efficiency will ever be the only or even the paramount goal. Defense contracts impose certain social goals on the defense industry labor force, like cultivating small, minority-owned, or disadvantaged subcontractors.

142. Although this issue was recently highlighted by defense industry leaders' complaints about their firms' stock prices during the late-1990s tech bubble, it is actually a timeworn issue for high-end engineering workers in the defense sector. See, for example, Claude Baum, *The System Builders: The Story of SDC* (Santa Monica, Calif.: System Development Corporation, 1981), pp. 129–31.

143. Private firms are sometimes accused of undervaluing research staff continuity in the face of investor pressure for short-term earnings. It is not clear why investors should be expected to make systematic mistakes in valuing research teams; they can simply discount future payoffs of research investment back to a net present value for comparing investments. In the 1990s, investors tended to overvalue the promise of technological progress, in the defense industry not least (expectations for which were briefly confused with those for the "dot-com" companies). Eugene Gholz, "Wall Street Lacks Realistic View of Defense Business," *Defense News*, December 20, 1999, p. 31.

144. Each of these sources of systems integration skill was cited in one or more interviews—usually in self-serving ways. That is, a systems-integration organization with close academic ties would emphasize the importance of access to basic scientific research to its work, while an organization with ties to a major defense production organization would emphasize production experience as a key underpinning of systems-integration skill.

145. Davis Dyer, *TRW: Pioneering Technology and Innovation since 1900* (Boston: Harvard Business School Press, 1998), pp. 225–39. Also, William L. Baldwin, *The Structure of the Defense Market 1955–1964* (Durham, N.C.: Duke University Press, 1967), pp. 45–46, 138–39. A similar situation led to the creation of MITRE. See John F. Jacobs, *The Sage Air Defense System: A Personal History* (Bedford, Mass.: MITRE Corporation, 1986), pp. 137–41.

146. Bruce L. R. Smith, *The Future of the Not-for-Profit Corporations*, P-3366 (Santa Monica, Calif.: RAND, May 1966), p. 18. Smith predicted that the FFRDC role would fade as the military improved its in-house technical capabilities. But for the reasons discussed in the text—and because the FFRDCs' success, which Smith underlines in his report, reduced the demand for in-house systems-integration capability—the military services never developed sufficient expertise to replace them. For-profit systems integration contractors (e.g., SAIC) have proven to be a bigger threat to the FFRDCs than resurgent government laboratories.

147. Johns Hopkins University APL is not technically an FFRDC at present (it was until 1977), but it remains a nonprofit systems-integration organization with a long-term contractual relationship with the U.S. Navy. Like an FFRDC, APL does not primarily engage in production, and it sometimes acts as the technical-direction agent on major naval systems contracts. For present purposes, APL can be grouped with MITRE and Aerospace as a systems-integration FFRDC, although it also has a strong research program, analogous to that of Lincoln Laboratory.

148. U.S. General Accounting Office, *Strategic Defense Initiative Program: Expert's Views on DoD's Organizational Options and Plans for SDI Technical Support*, GAO/NSIAD-87-43 (Washington, D.C.: November 1986), p. 4.

149. U.S. General Accounting Office, *Federally Funded R&D Centers: Issues Relating to the Management of DoD-Sponsored Centers*, GAO/NSIAD-96-112 (Washington, D.C.: August 1996), pp. 5–6; U.S. Congress, Office of Technology Assessment [hereafter OTA], *A History of the Department of Defense Federally Funded Research and Development Centers*, OTA-BP-ISS-157 (Washington, D.C.: U.S. Government Printing Office, June 1995), pp. 28–33. SAIC specifically acknowledges the technical skills of FFRDCs and actually tried to purchase Aerospace Corporation in 1996—claiming that it could maintain that organization's skills while adding efficiency due to the profit motive. Air Force resistance blocked this controversial move; many scientists at Aerospace were also skeptical of the acquisition and now report that they would have considered leaving the company if the SAIC deal had gone though. See John Mintz, "Air Force Halts Merger of 2 Companies," *Washington Post*, November 16, 1996, p. D1.

150. Some involved in these congressional decisions believe that the perceived high cost of FFRDCs was the crucial issue in establishing these limits; others see the effects of a lingering controversy over missile defense. The most recent proposal to establish a new FFRDC would have created a Strategic Defense Initiative Institute to support the missile defense effort.

151. SYNTEK, for example, has benefited by hiring a number of technical experts who had gained experience working in military laboratories at a time (in the 1960s and 1970s) when they had stronger roles in architecture definition. SYNTEK executives fear that their skills will be hard to maintain in future generations of technical staff. Author interviews, September 2000.

152. The SEI has begun to develop a new Capabilities Maturity Model to evaluate "integration" skills. At the direction of OSD, it is trying to apply software systems engineering procedures to software-hardware integration. The goal is to develop best-practice methodologies for reducing the rate of failures in complex projects. Even this ongoing broadening of the SEI's research remains at a "lower" level than the overall system-of-systems integration that is a key initial step in transformation.

153. In interviews, several respondents noted that the CMM-I project was causing tension between the SEI and MITRE, as both clamored for the attention of key customers at the Air Force Electronic Systems Command at Hanscom Air Force Base.

154. Thomas P. Hughes, *Rescuing Prometheus: Four Monumental Projects That Changed the Modern World* (New York: Vintage Books, 1998).

155. POET substantially outlived the particular "Phase One" referred to by its title. Recently, the reorganization of the BMDO into the Missile Defense Agency has been accompanied by the creation of a "National Team" to provide technical support and systems integration for missile defense. The National Team involves prime contractors that produce platforms—specifically including platforms that will be deployed as part of the tiered missile-defense system of systems.

156. Author interview, August 2001.

157. Author interview, July 2002.

158. Jacobs, p. 131; Hughes, p. 62; Baum, pp. 38–39.

159. A similar idea was proposed to provide technical support to the missile defense program—either personnel from established FFRDCs would have been reassigned to the new SDI Institute, or a new division of one of the established FFRDCs would have been created. This approach was rejected in favor of the POET, arguably because the new FFRDC approach was perceived as too slow to set up and too costly. Others suggest that the SDII proposal was blocked by political opponents of missile defense, who hoped to hamstring the effort by denying high-quality technical advice to the Strategic Defense Initiative Office. See Donald Baucom, "The Rise and Fall of the SDI Institute: A Case Study of the Management of the Strategic Defense Initiative," incomplete draft, August 1998.

160. Thomas L. McNaugher, *New Weapons, Old Politics: America's Military Procurement Muddle* (Washington, D.C.: Brookings Institution, 1989), pp. 3–12.

161. Harvey Sapolsky, "Myth and Reality in Project Planning and Control," in *Macro-Engineering and the Future*, ed. F. Davidson and C. Lawrence Meadow (Boulder, Colo.: Westview Press, 1982), pp. 173–82. On rare occasions, oversight officials or firms have been known to falsify reports, but those cases are truly the exception rather than the rule. Robert Wall, "V-22 Support Fades amid Accidents, Accusations, Probes," *Aviation Week and Space Technology,* January 29, 2001, p. 28.

162. Cindy Williams, "Holding the Line on Infrastructure Spending," in *Holding the Line*, pp. 55–77.

163. Harvey M. Sapolsky, *The Polaris System Development: Bureaucratic and Programmatic Success in Government* (Cambridge, Mass.: Harvard University Press, 1972).

164. Conflicts among those tasks have been barriers to the successful application of the systems approach outside of the acquisition environment. Stephen P. Rosen, "Systems Analysis and the Quest for Rational Defense," *Public Interest*, no. 76 (Summer 1984), pp. 3–17.

165. For a general discussion of this form of organizational behavior, see Jeffrey Pfeffer, "A Resource Dependence Perspective on Intercorporate Relations," in Mark S. Mizruchi and Michael Schwartz, eds., *Intercorporate Relations: The Structural Analysis of Business* (New York: Cambridge University Press, 1987).

166. William Rogerson, "Incentives in Defense Contracting," paper presented at the MIT Security Studies Program, October 1998; Thomas L. McNaugher, "Weapons Procurement: The Futility of Reform," in *America's Defense*, ed. M. Mandelbaum (New York: Holmes and Meier, 1989), pp. 68–112.

167. Gholz and Sapolsky, "Restructuring the U.S. Defense Industry."

168. Office of Technology Assessment, *History of Department of Defense Federally Funded Research and Development Centers.*

169. Phil Balisle and Tom Bush, "CEC Provides Theater Air Dominance," U.S. Naval Institute *Proceedings* 128, no. 5 (May 2002), pp. 60–62 and the responses in the July and August 2002 issues of *Proceedings.*

170. Author interviews, May 2002; Gopal Ratnam, "U.S. Navy to Set New CEC Requirements," *Defense News*, July 22–28, 2002, p. 44.

171. McNaugher, "Weapons Procurement"; Ethan McKinney, Eugene Gholz, and Harvey M. Sapolsky, *Acquisition Reform*, MIT Lean Aircraft Initiative Policy Working Group, Working Paper 1, 1994.

172. Carl H. Builder, *The Masks of War* (Baltimore: Johns Hopkins University Press, 1989).

173. This requirement is another reason why it is difficult for government agencies to perform systems integration in-house: subordinate project managers in the systems commands might not risk criticizing their bosses' (or their bosses') preferred programs. OTA, *History of Department of Defense Federally Funded Research and Development Centers*, p. 5. Quasi-public FFRDCs face similar pressure not to criticize their customers too much, but their support and promotion prospects do not come through as direct a chain of command from the potential targets of their technical advice. The position of

for-profit systems integration houses is similar to that of the FFRDCs; they may be more responsive to short-term budget pressures from sponsoring organizations than FFRDCs are, but they may have more independence to seek alternative customers if a relationship with a particular contracting command temporarily sours.

174. James Q. Wilson, *Bureaucracy: What Government Agencies Do and Why They Do It* (New York: Basic Books, 1989).

175. For example, Lockheed Martin has a large systems-integration group in Valley Forge, Pennsylvania, with specific expertise in satellites and intelligence collection. Lockheed Martin, of course, would need to keep some proprietary systems integration capability, even if it were clear that the Navy did not plan to delegate high-level systems integration/technical decision making to the production prime contractors. Each member of the production defense industrial base would then have to make a business decision about what level of in-house funding to allot to SI, given that the main institutional home of that core competency would be outside the production industrial base.

176. Author interview, December 2000.

177. According to General Dynamics, its acquisition of Newport News Shipyard would have resulted in "savings of $2 billion or more over 10 years." Christopher J. Castelli, "GD, Newport News Execs Say Merger Savings Akin to 1999 Projections," *Inside the Navy*, April 30, 2001, p. 6.

178. See Williams, "Holding the Line on Infrastructure Spending," pp. 55–77.

179. Gail Kaufman and Gopal Ratnam, "The Search for an Affordable UAV," *Defense News*, September 2–8, 2002, pp. 1 and 10.

180. Rosen, *Winning the Next War*, pp. 22, 250.

Selected Bibliography

BOOKS

Alberts, David S., John J. Garstka, and Frederick P. Stein. *Network Centric Warfare: Developing and Leveraging Information Superiority*. 2nd ed. Washington, D.C.: C4ISR Cooperative Research Program, 1999.

Alic, John A., L. M. Branscomb, A. B. Carter, and G. L. Epstein. *Beyond Spinoff: Military and Commercial Technologies in a Changing World*. Boston: Harvard Business School Press, 1992.

Arquilla, John, and David Ronfeldt, eds. *Athena's Camp: Preparing for Conflict in the Information Age*. Santa Monica, Calif.: RAND, 1997.

Baldwin, William L. *The Structure of the Defense Market 1955–1964*. Durham, N.C.: Duke University Press, 1967.

Bell, Daniel. *The Coming of Post-Industrial Society: A Venture in Social Forecasting*. New York: Basic Books, 1999.

Binnendijk, Hans, ed. *Transforming America's Military*. Washington, D.C.: National Defense University, 2002.

Builder, Carl H. *The Masks of War: American Military Styles in Strategy and Analysis*. Baltimore: Johns Hopkins University Press, 1989.

Castells, Manuel. *The Rise of the Network Society*. 2nd ed. Oxford: Blackwell, 2000.

Christensen, Clayton M. *The Innovator's Dilemma: When New Technologies Cause Great Firms to Fail*. Boston: Harvard Business School Press, 1997.

Committee on Network-Centric Naval Forces, Naval Studies Board. *Network-Centric Naval Forces: A Transition Strategy for Enhancing Operational Capabilities*. Washington, D.C.: National Academy Press, 2000.

Dyer, Davis. *TRW: Pioneering Technology and Innovation since 1900*. Boston: Harvard Business School Press, 1998.

Evangelista, Matthew. *Innovation and the Arms Race: How the United States and the Soviet Union Develop New Military Technologies*. Ithaca, N.Y.: Cornell University Press, 1988.

Friedberg, Aaron L. *In the Shadow of the Garrison State: America's Anti-Statism and Its Cold War Strategy*. Princeton, N.J.: Princeton University Press, 2000.

Friedman, Norman. *Seapower and Space: From the Dawn of the Missile Age to Net-Centric Warfare*. Annapolis, Md.: Naval Institute Press, 2000.

Friedman, Thomas L. *The Lexus and the Olive Tree: Understanding Globalization*. New York: Farrar, Straus, and Giroux, 1999.

Gansler, Jacques S. *Defense Conversion: Transforming the Arsenal of Democracy* Cambridge, Mass.: MIT Press, 1995.

Gansler, Jacques S. *The Defense Industry*. Cambridge, Mass.: MIT Press, 1980.

Gleick, James. *Faster: The Acceleration of Just About Everything*. New York: Pantheon, 1999.

Hughes, Thomas P. *Rescuing Prometheus: Four Monumental Projects That Changed the Modern World*. New York: Vintage Books, 1998.

Kier, Elizabeth. *Imagining War: French and British Military Doctrine between the Wars*. Princeton, N.J.: Princeton University Press, 1997.

Koistinen, Paul A. C. *Beating Plowshares into Swords: The Political Economy of American Warfare, 1606–1865*. Lawrence: University Press of Kansas, 1996.

———. *Mobilizing for Modern War: The Political Economy of American Warfare, 1865–1919*. Lawrence: University Press of Kansas, 1997.

———. *Planning War, Pursuing Peace: The Political Economy of American Warfare, 1920–1939*. Lawrence: University Press of Kansas, 1998.

Markusen, Ann R., and Sean S. Costigan, eds. *Arming the Future: A Defense Industry for the 21st Century*. New York: Council on Foreign Relations, 1999.

McNaugher, Thomas L. *New Weapons, Old Politics: America's Military Procurement Muddle.* Washington, D.C.: Brookings Institution, 1989.

Munson, Kenneth, ed. *Jane's Unmanned Aerial Vehicles and Targets*, no. 17. Coulsdon, Surrey, U.K., and Alexandria, Va.: Jane's Information Group, December 2001.

O'Hanlon, Michael. *Technological Change and the Future of Warfare.* Washington, D.C.: Brookings Institution, 2000.

Owens, William A., with Edward Offley. *Lifting the Fog of War.* New York: Farrar, Straus and Giroux, 2000.

Posen, Barry R. *The Sources of Military Doctrine: France, Britain, and Germany between the Wars.* Ithaca, N.Y.: Cornell University Press, 1984.

Rosen, Stephen Peter. *Winning the Next War: Innovation and the Modern Military.* Ithaca, N.Y.: Cornell University Press, 1991.

Ross, Andrew L. *The Political Economy of Defense: Issues and Perspectives.* Westport, Conn.: Greenwood Press, 1991.

Sapolsky, Harvey M. *The Polaris System Development: Bureaucratic and Programmatic Success in Government.* Cambridge, Mass.: Harvard University Press, 1972.

———. *Science and the Navy: The History of the Office of Naval Research.* Princeton, N.J.: Princeton University Press, 1990.

Toffler, Alvin. *Powershift: Knowledge, Wealth, and Violence at the Edge of the 21st Century.* New York: Bantam Books, 1990.

———. *The Third Wave.* New York: William Morrow, 1980.

Toffler, Alvin, and Heidi Toffler. *Creating a New Civilization: The Politics of the Third Wave.* Atlanta: Turner, 1995.

———. *War and Anti-War: Making Sense of Today's Global Chaos.* New York: Warner, 1993.

Utterback, James M. *Mastering the Dynamics of Innovation.* Boston: Harvard Business School Press, 1994.

Von Hippel, Eric. *The Sources of Innovation.* New York: Oxford University Press, 1988.

Williams, Cindy, ed. *Holding the Line: U.S. Defense Alternatives for the Early 21st Century.* Cambridge, Mass.: MIT Press, 2001.

Wilson, James Q. *Bureaucracy: What Government Agencies Do and Why They Do It.* New York: Basic Books, 1989.

ARTICLES and CHAPTERS

Balisle, Phil, and Captain Tom Bush. "CEC Provides Theater Air Dominance." U.S. Naval Institute *Proceedings* 128, no. 5 (May 2002): 60–62.

Barnett, Thomas P. M. "The Seven Deadly Sins of Network-Centric Warfare." U.S. Naval Institute *Proceedings* (January 1999): 36–39.

Bower, Joseph L., and Clayton M. Christensen. "Disruptive Technologies: Catching the Wave." *Harvard Business Review* (January–February 1995): 43–53.

Bracken, Paul. "The Military after Next." *Washington Quarterly* 16, no. 4 (Autumn 1993): 157–74.

Cebrowski, Vice Admiral Arthur K., and John J. Garstka. "Network-Centric Warfare: Its Origin and Future." U.S. Naval Institute *Proceedings* (January 1998): 28–35.

Christensen, Clayton M. "The Rules of Innovation." *Technology Review* (June 2002): 33–38.

Christensen, Clayton, Thomas Craig, and Stuart Hart. "The Great Disruption." *Foreign Affairs* 80, no. 2 (March–April 2001): 80–95.

Christensen, Clayton M., Mark V. Johnson, and Darrell K. Rigby. "Foundations for Growth: How to Identify and Build Disruptive New Businesses." *MIT Sloan Management Review* 43, no. 3 (Spring 2002): 22–31.

Christensen, Clayton M., and Michael Overdorf. "Meeting the Challenge of Disruptive Change." *Harvard Business Review* (March–April 2000): 67–76.

Christensen, Clayton M., and Richard S. Tedlow. "Patterns of Disruption in Retailing." *Harvard Business Review* (January–February 2000): 42–45.

Cohen, Eliot A. "A Revolution in Warfare." *Foreign Affairs* 75, no. 2 (March–April, 1996): 37–54.

Cook, Nick. "Network-Centric Warfare: The New Face of C4I." *Interavia* (February 2001): 37–39.

Cote, Owen R., Jr. "Buying '... From the Sea': A Defense Budget for a Maritime Strategy." In *Holding the Line: U.S. Defense Alternatives for the Early 21st Century*. Edited by Cindy Williams. Cambridge, Mass.: MIT Press, 2001.

Dawson, J. Cutler, Jr., James M. Fordice, and Gregory M. Harris. "The IT-21 Advantage." U.S. Naval Institute *Proceedings* (December 1999): 28–32.

DeMarines, Victor A., with David Lehman and John Quilty. "Exploiting the Internet Revolution." In *Keeping the Edge: Managing Defense for the Future*. Edited by Ashton B. Carter and John P. White. Cambridge, Mass., and Stanford, Calif.: Preventive Defense Project, 2000.

Dombrowski, Peter. "The Globalization of the Defense Sector? Naval Industrial Cases and Issues." In *Globalization and Maritime Power*. Edited by Sam J. Tangredi. Washington, D.C.: National Defense University Press, 2002.

Dombrowski, Peter J., Eugene Gholz, and Andrew L. Ross. "Selling Military Transformation: The Defense Industry and Innovation." *Orbis*, 46, no. 3 (Summer 2002): 523–36.

Gholz, Eugene. "The Irrelevance of International Defense Mergers." *Breakthroughs* 9, no. 1 (Spring 2000): 3–11.

Gholz, Eugene, and Harvey Sapolsky. "Restructuring the U.S. Defense Industry." *International Security* 24, no. 3 (Winter 1999–2000): 5–51.

Glade, David B., II. "Unmanned Aerial Vehicles." In *The Technological Arsenal: Emerging Defense Capabilities*. Edited by William C. Martel. Washington, D.C.: Smithsonian Institution Press, 2001.

Goodman, Glenn W., Jr. "Manned-Unmanned Synergy: US Army's UAV-Related Efforts Gain Momentum." *Armed Forces Journal International* (July 2002): 56–61.

Hamilton, Rear Admiral Charles, USN, and Rear Admiral Donald Loren, USN. "It's All in the Family." U.S. Naval Institute *Proceedings* (August 2002): 68–70.

Harknett, Richard J., and the JCISS Study Group. "The Risks of a Networked Military." *Orbis* 44, no. 1 (Winter 2000): 127–43.

Holland, John H. "What Is to Come and How to Predict It." In *The Next Fifty Years: Science in the First Half of the Twenty-first Century*. Edited by John Brockman. New York: Vintage Books, 2002.

Hoffman, Frank. "Goldwater-Nichols after a Decade." In *The Emerging Strategic Environment: Challenges of the Twenty-first Century*. Edited by Willamson Murray Westport, Conn.: Praeger, 1999.

Hone, Thomas C. "Force Planning Cycles: The Modern Navy as an Illustrative Case of a Frustrating Trend." *Defense Analysis* 9, no. 1 (April 1993): 31–42.

Kennedy, Harold. "U.S. Services Test Aussie-Built Catamaran." *National Defense* (April 2002): 30–31.

King, James A. "Stealth Means Survivability." U.S. Naval Institute *Proceedings* (December 2001): 80–83.

Markusen, Ann. "The Rise of World Weapons." *Foreign Policy*, no. 114 (Spring 1999): 40–51.

McNaugher, Thomas L. "Weapons Procurement: The Futility of Reform." In *America's Defense*. Edited by M. Mandelbaum. New York: Holmes and Meier, 1989.

Natter, Admiral Robert J., U.S. Navy. "Meeting the Need for Speed." U.S. Naval Institute *Proceedings* (June 2002): 65–67.

Nardulli, Bruce R., and Thomas L. McNaugher. "The Army: Toward the Objective Force." In *Transforming America's Military*. Edited by Hans Binnendijk. Washington, D.C.: National Defense University, 2002.

Nye, Joseph S., Jr. and William A. Owens. "America's Information Edge." *Foreign Affairs* 75, no. 2 (March–April 1996): 20–36.

Ochmanek, David. "The Air Force: The Next Round." In *Transforming America's Military*. Edited by Hans Binnendijk. Washington, D.C.: National Defense University, 2002.

Oden, Michael. "Cashing In, Cashing Out, and Converting: Restructuring of the Defense Industrial Base in the 1990s." In *Arming the Future: A Defense Industry for the 21st Century*. Edited by Ann R. Markusen and Sean S. Costigan. New York: Council on Foreign Relations Press, 1999.

O'Neil, William D. "The Naval Services: Network-Centric Warfare." In *Transforming America's Military*. Edited by Hans Binnendijk. Washington, D.C.: National Defense University, 2002.

Owens, William A. "The Emerging System of Systems." U.S. Naval Institute *Proceedings* (May 1995): 36–39.

Pierce, Captain Terry C. "Jointness Is Killing Naval Innovation." U.S. Naval Institute *Proceedings* (October 2001): 68–71.

Rauch, Jonathan. "The New Old Economy: Oil, Computers, and the Reinvention of the Earth." *Atlantic Monthly* (January 2001): 35–49.

Roos, John G. "An All-Encompassing Grid." *Armed Forces Journal International* (January 2001): 26–35.

Rosen, Stephen P. "Systems Analysis and the Quest for Rational Defense." *Public Interest*, no. 76 (Summer 1984): 3–17.

Ross, Andrew L. "Defense Industry Globalization: Contrarian Observations." In *Defense Industry Globalization*. Washington, D.C.: Atlantic Council of the United States, February 2002.

———. "The Dynamics of Military Technology." In *Building a New Global Order: Emerging Trends in International Security*. Edited by David Dewitt, David Haglund, and John Kirton. Toronto: Oxford University Press, 1993.

Saalfeld, Fred E., and John F. Petrik. "Disruptive Technologies: A Concept for Moving Innovative Military Technologies Rapidly to Warfighters." *Armed Forces Journal International* (May 2001): 48–52.

Sapolsky, Harvey M. "Myth and Reality in Project Planning and Control." In *Macro-Engineering and the Future*. Edited by F. Davidson and C. Lawrence Meadow. Boulder, Colo.: Westview Press, 1982.

———. "Equipping the Armed Forces." In *National Security and the U.S. Constitution*. Edited by George Edwards and W. Earl Walker. Baltimore: Johns Hopkins University Press, 1988.

Sapolsky, Harvey, and Eugene Gholz. "Eliminating Excess Defense Production." *Issues in Science and Technology* 13, no. 2 (Winter 1996–97): 65–70.

———. "The Defense Industry's New Cycle." *Regulation* 24, no. 3 (Winter 2001–2002): 44–49.

Sapolsky, Harvey M., Eugene Gholz, and Allen Kaufman. "Security Lessons from the Cold War." *Foreign Affairs* 78, no. 4 (July–August, 1999): 77–89.

Smith, Edward P. "Network-Centric Warfare: What's the Point?" *Naval War College Review* 54, no. 1 (Winter 2001): 59–75.

Weiner, Sharon. "The Politics of Resource Allocation in the Post–Cold War Pentagon." *Security Studies* 5, no. 4 (Summer 1996): 125–42.

Wood, Geoffrey. "The Rise of Unconventional Naval Platforms." *Military Technology* (May 2002): 58–63.

REPORTS/MONOGRAPHS

Axelson, Mattias, and Andrew James. *The Defence Industry and Globalisation: Challenging Traditional Structures*, FOA-R-00-01698-170-SE. Stockholm: Defense Research Establishment, December 2000.

Baum, Claude. *The System Builders: The Story of SDC*. Santa Monica, Calif.: System Development Corporation, 1981.

Birkler, John, Giles Smith, Glenn A. Kent, and Robert V. Johnson. *An Acquisition Strategy, Process, and Organization for Innovative Systems.* Santa Monica, Calif.: RAND, 2000.

Bischak, Greg. *US Conversion after the Cold War, 1990–1997: Lessons for Forging a New Conversion Policy*, brief 9. Bonn: Bonn International Center for Conversion, July 1997.

Bolkom, Christopher. *Air Force Transformation and Modernization: Overview and Issues for Congress*, RS20787. Washington, D.C.: Congressional Research Service, Library of Congress, 1 June 2001.

Bruner, Edward F. *Army Transformation and Modernization: Overview and Issues for Congress*, RS20787. Washington, D.C.: Congressional Research Service, Library of Congress, 4 April 2001.

Cote, Owen R., Jr. *The Politics of Innovative Military Doctrine: The U.S. Navy and Fleet Ballistic Missiles.* Ph.D. dissertation. Cambridge: Massachusetts Institute of Technology, 1995.

Defense Science Board Task Force. *Preserving a Healthy and Competitive U.S. Defense Industry to Ensure Our Future National Security*, Final Briefing, November 2000.

Ehrhard, Thomas P. *Unmanned Aerial Vehicles in the United States Armed Services: A Comparative Study of Weapon System Innovation.* Ph.D. dissertation. Baltimore, Md.: Johns Hopkins University, 2001.

Harbison, John R., Thomas S. Moorman, Jr., Michael W. Jones, and Jikun Kim. *U.S. Defense Industry: An Agenda for Change.* Booz Allen Hamilton, 2000.

Jacobs, John F. *The Sage Air Defense System: A Personal History.* Bedford, Mass.: MITRE Corporation, 1986.

Labs, Eric. *Options for Enhancing the Department of Defense's Aerial Vehicles Programs.* Washington, D.C.: Congressional Budget Office, September 1998.

Leonard, Robert S., Jeffrey A. Drezner, and Geoffrey Sommer. *The Arsenal Ship: Acquisition Process Experience.* Santa Monica, Calif.: RAND, 1999.

Naval Transformation Roadmap: Power and Access . . . from the Sea. Washington, D.C.: Department of the Navy, 2002.

Navy Warfare Development Command. *Network Centric Operations: A Capstone Concept for Naval Operations in the Information Age.* Newport, R.I.: draft dated 6/19/01, available at www.nwdc.navy.mil/Concepts/capstone _concept.asp.

Odell, Robert, Bruce Wald, Lyntis Beard, with Jack Batzler and Michael Loescher. *Taking Forward the Navy's Network-Centric Warfare Concept: Final Report*, CRM 99-42.10. Alexandria, Va.: Center for Naval Analyses, May 1999.

Office of Technology Assessment, U.S. Congress. *A History of the Department of Defense Federally Funded Research and Development Centers*, OTA-BP-ISS-157. Washington, D.C.: U.S. Government Printing Office, June 1995.

O'Rourke, Ronald. *Coast Guard Deepwater Program: Background and Issues for Congress*, RS21019. Washington, D.C.: Congressional Research Service, Library of Congress, 23 May 2002.

———. *Navy CVNX Aircraft Carrier Program: Background and Issues for Congress*, RS20643. Washington, D.C.: Congressional Research Service, Library of Congress, 23 May 2002.

———. *Navy DD(X) Future Surface Combatant Program: Background and Issues for Congress*, RS21059. Washington, D.C.: Congressional Research Service, Library of Congress, 10 May 2002.

———. *Navy Major Shipbuilder Ownership Consolidation: Issues for Congress*, RL3051. Washington, D.C.: Congressional Research Service, Library of Congress, 7 July 1999.

———. *Navy Network-Centric Warfare Concept: Key Programs and Issues for Congress*, RS20557. Washington, D.C.: Congressional Research Service, Library of Congress, 6 June 2001.

———. *Navy Shipbuilding: Proposed Mergers Involving Newport News Shipbuilding—Issues for Congress*, RL30969. Washington, D.C.: Congressional Research Service, 22 May 2001.

———. *Navy Shipbuilding: Recent Shipyard Mergers—Background and Issues for Congress*, RL31400. Washington, D.C.: Congressional Research Service, 3 May 2002.

Owens, William A. "The Emerging U.S. System-of-Systems." Strategic Forum 63. Washington, D.C.: Institute for National Strategic Studies, National Defense University, February 1996.

U.S. Department of Commerce. *National Security Assessment of the U.S. Shipbuilding and Repair Industry*. Washington, D.C.: U.S. Department of Commerce, Bureau of export Administration, Office of Strategic Industries and Economic Security, May 2001.

U.S. Department of Defense. *Network Centric Warfare*. Washington, D.C.: 27 July 2001, available at www.c3i.osd.mil/NCW.

U.S. General Accounting Office. *Federally Funded R&D Centers: Issues Relating to the Management of DoD-Sponsored Centers*, GAO/NSIAD-96-112. Washington, D.C.: August 1996.

———. *Strategic Defense Initiative Program: Expert's Views on DoD's Organizational Options and Plans for SDI Technical Support*, GAO/NSIAD-87-43. Washington, D.C.: November 1986.

Selected Briefings

American Institute of Engineers, Network Centric Warfare Conference

Armed Forces Communications & Electronics Association, Lexington-Concord Chapter, New Horizons Symposium

Armed Forces Communications & Electronics Association, Newport Chapter

Booz Allen Hamilton

Eurasian Studies Group, Naval War College

Government Electronics and Information Technology Association

National Conference of Editorial Writers

Office of Naval Research, Defense Manufacturers Conference 2003

Office of Naval Research, ShipTech 2003

President, Naval War College

RAND Washington Office, monthly Navy Luncheon

Rhode Island Economic Development Council

Secretary of the Navy Gordon R. England

United Defense LP

World Affairs Council of Boston

About the Authors

DR. PETER J. DOMBROWSKI is an associate professor in the Strategic Research Department of the U.S. Naval War College. Prior to assuming this position, he was an associate professor of political science at Iowa State University. He has published research on various topics in national security strategy, American foreign policy, international relations, and international political economy. In 1996, the University of Pittsburgh Press published his *Policy Responses to the Globalization of American Banks*. Recent articles include "Against Pre-emptive Strikes," posted on *The Information Technology, War and Peace Project* website, http://www.watsoninstitute.org/infopeace/911/new; "Selling Military Transformation: The Defense Industry and Innovation," in *Orbis*; "Arguments for a Renewed RMA Debate," in *National Security Studies Quarterly*; and "Military Relations with Humanitarian Organizations in Complex Emergencies," in *Global Governance*. Dr. Dombrowski is now serving a five-year term as co-editor of the journal *International Studies Quarterly*. He holds a B.A. degree from Williams College and an M.A. and Ph.D. from the University of Maryland.

DR. EUGENE GHOLZ is an assistant professor at the Patterson School of Diplomacy and International Commerce at the University of Kentucky. He takes the lead role in the international commerce part of the Patterson School curriculum, which trains master's students for government, nongovernmental organization, and private-sector jobs that emphasize skills in international political economy. Professor Gholz teaches courses on globalization, economic statecraft, and defense statecraft. His research concerns how the government decides what weapons to buy, how and when to stimulate innovation, and how to manage high-tech business-government relations from both business and government perspectives. He has published numerous articles in leading journals of international and national security affairs, including *International Security, Foreign Affairs, Security Studies, Orbis*, and *Issues in Science and Technology*. Dr. Gholz taught previously in George Mason University's International Commerce and Policy Program. Prior to that, he was a national security fellow at Harvard University's Olin Institute of Strategic Studies. He received his Ph.D. from the Massachusetts Institute of Technology's Department of Political Science.

DR. ANDREW L. ROSS is professor of strategic studies and director of studies in the Strategic Research Department of the U.S. Naval War College's Center for Naval Warfare Studies. He served as director of the Naval War College's project on "Military Transformation and the Defense Industry after Next." During the 2001–2002 academic year,

Dr. Ross was the acting director of the College's Advanced Research Program and a co-leader of the College's Strategy Task Group, one of four task groups established to support OpNav in the global war on terror. His work on grand strategy, defense planning, regional security, weapons proliferation, the international arms market, defense industrialization, and security and development has appeared in numerous journals and books. Professor Ross is the editor of *The Political Economy of Defense: Issues and Perspectives* (1991) and co-editor of three editions of *Strategy and Force Planning* (1995, 1997, 2000). Dr. Ross has held research fellowships at Cornell, Princeton, Harvard, the University of Illinois, and the Naval War College and has taught in the Political Science Departments of the University of Illinois and the University of Kentucky. He earned his M.A. and Ph.D. at Cornell University and his B.A., summa cum laude, at American University.

Titles in the Series

"Are We Beasts?" Churchill and the Moral Question of World War II "Area Bombing," by Christopher C. Harmon (December 1991).

Toward a Pax Universalis: A Historical Critique of the National Military Strategy for the 1990s, by Lieutenant Colonel Gary W. Anderson, U.S. Marine Corps (April 1992).

The "New" Law of the Sea and the Law of Armed Conflict at Sea, by Horace B. Robertson, Jr. (October 1992).

Global War Game: The First Five Years, by Bud Hay and Bob Gile (June 1993).

Beyond Mahan: A Proposal for a U.S. Naval Strategy in the Twenty-First Century, by Colonel Gary W. Anderson, U.S. Marine Corps (August 1993).

The Burden of Trafalgar: Decisive Battle and Naval Strategic Expectations on the Eve of the First World War, by Jan S. Breemer (October 1993).

Mission in the East: The Building of an Army in a Democracy in the New German States, by Colonel Mark E. Victorson, U.S. Army (June 1994).

Physics and Metaphysics of Deterrence: The British Approach, by Myron A. Greenberg (December 1994).

A Doctrine Reader: The Navies of the United States, Great Britain, France, Italy, and Spain, by James J. Tritten and Vice Admiral Luigi Donolo, Italian Navy (Retired) (December 1995).

Chaos Theory: The Essentials for Military Applications, by Major Glenn E. James, U.S. Air Force (October 1996).

The International Legal Ramifications of United States Counter-Proliferation Strategy: Problems and Prospects, by Frank Gibson Goldman (April 1997).

What Color Helmet? Reforming Security Council Peacekeeping Mandates, by Myron H. Nordquist (August 1997).

Sailing New Seas, by Admiral J. Paul Reason, U.S. Navy, Commander-in-Chief, U.S. Atlantic Fleet, with David G. Freymann (March 1998).

Theater Ballistic Missile Defense from the Sea: Issues for the Maritime Component Commander, by Commander Charles C. Swicker, U.S. Navy (August 1998).

International Law and Naval War: The Effect of Marine Safety and Pollution Conventions during International Armed Conflict, by Dr. Sonja Ann Jozef Boelaert-Suominen (December 2000).

The Third Battle: Innovation in the U.S. Navy's Silent Cold War Struggle with Soviet Submarines, by Owen R. Cote, Jr. (2003).

The Limits of Transformation: Officer Attitudes toward the Revolution in Military Affairs, by Thomas G. Mahnken and James R. FitzSimonds (2003).

Forthcoming in 2003

Global War Game: Second Series, 1984–1988, by Robert H. Gile.

Forthcoming in 2004

The Evolution of the U.S. Navy's Maritime Strategy, 1977–1986, by John B. Hattendorf.